LOOKING BEYOND THE RUNWAY

To my wife, Carolyn

Looking Beyond the Runway

Airlines Innovating with Best Practices while Facing Realities

NAWAL K. TANEJA

LONDON AND NEW YORK

First published 2010 by Ashgate Publishing

Published 2016 by Routledge
2 Park Square, Milton Park, Abingdon, Oxon OX14 4RN
711 Third Avenue, New York, NY 10017, USA

Routledge is an imprint of the Taylor & Francis Group, an informa business

British Library Cataloguing in Publication Data
Taneja, Nawal K.
Looking beyond the runway : airlines innovating with best practices while facing realities.
1. Airlines--Management. 2. Organizational change--Case studies. 3. Strategic planning--Case studies.
I. Title
387.7'068-dc22

978-1-4094-0099-8 (hbk)

Library of Congress Cataloging-in-Publication Data
Taneja, Nawal K.
Looking beyond the runway: airlines innovating with best practices while facing realities / by Nawal K. Taneja.
p. cm.
Includes index.
ISBN 978-1-4094-0099-8 (hardback)
1. Airlines. 2. Airlines--Customer services. 3. Automobile industry and trade.
4. Aeronautics, Commercial. 5. Consumer satisfaction. I. Title.
HE9776.T365 2010
387.7068'4--dc22

2009050842

ISBN: 9781409400998 (hbk)

Contents

List of Figures

List of Tables

Preface

Looking Beyond the Runway is the sixth book in a series written for, and at the encouragement of, practitioners in the global airline industry.[1] The current chaos, led by the unusual downturn that started at the end of 2007, is proving to be a serious challenge for the global airline industry. While most airline managements are developing strategies just to survive through this chaos, a few airline managements see it as a great opportunity for positioning their airlines to achieve profitable and sustainable growth. This latter group of executives is neither resisting change, nor blaming the industry's constraints as the major cause of their inability to change. On the contrary, this group is welcoming the opportunity to learn from the failures and successes of others (businesses, both within and outside the global industry) to renovate their ageing business models around the constraints that are inherent in the airline industry.

The content of this book is divided into seven chapters to address the aforementioned challenges and opportunities.

- Chapter 1, *Outlining the Chaos, Evolving Strategies, and the New Normal*, provides an overview of the chaos facing the global airline industry, a brief commentary on the strategies being implemented to deal with the chaos (by the defensive and the offensive groups of airlines), and a hypothesis that there will be a new normal, for, at least those airlines based in North America, parts of Europe, and parts of Asia. In other words, the new normal may be of less concern to airlines based in such regions as the Middle East, and parts of Africa, Latin America, and Asia. While different regions are expected to be impacted to different degrees by different factors, the expansion of low-cost carriers is expected to be the single factor that will have an enormous influence on legacy carriers based in every region.

- Chapter 2, *Learning from Other Struggles: The Auto Industry*, deals with the experiences of the auto industry. Given the similarities between the two industries, particularly with respect to the Detroit Three, there are significant insights for airlines, especially with some based in the US. Examples include the cost of resisting change, focusing on market share, and lacking business agility. On the positive side are the insights from the success of Toyota, relating to the culture of experimentation.

- Chapter 3, *Learning from Other Successes: The Customer Experience Industry*, has some insights for airlines. Given the poor experience often provided by the airline industry, even though some elements of the travel process affecting customer service are outside the control of airlines, there is ample room for airlines to learn and improve. In the US, for example, responding to and attracting the emerging generation (the Millennials) as customers will require quite different approaches than in the past. In other parts of the world, the customer experience attributes desired may well be very different in light of significant differences in the stage and role of technology, channels of distribution, and the relationship of airline-communicated customer expectations and service delivered. The key insight for airlines from the customer experience-focused businesses (ranging from Zappos shoes to the Ritz-Carlton) is still valid worldwide, namely, their obsession with processes to select, equip, and retain employees (for example, through customer engagement).

- Chapter 4, *Innovating around Airline Realities*, deals with innovation, from incremental to disruptive, exemplified by the business model of Cirque du Soleil outside the airline industry and Ryanair within the airline industry. This chapter describes four major drivers of innovation and several case studies of companies who have excelled in the area of innovation (ranging from Tata Motors to Nintendo to Zipcar). While there are inherent constraints under which they operate, airlines must look for ways to innovate around these constraints as the truly binding constraints are not likely to be lifted by governments in the near future.

- Chapter 5, *Firing on All Cylinders to Stay Ahead*, summarizes a few key points for embracing and responding to the new normal through elevated airline business intelligence to sharpen focus on the core business and related businesses, business analytics to evaluate strategies relating to customers and competitors, and business agility to stay ahead of competition and build maneuverability. The chapter also provides some insights on value and pricing (including some best practices) as well as an example of a potential opportunity in pricing airline services based on merchandising models. Admitted, that the retail industry is not only very diverse, but also that the airline business is becoming increasingly commoditized; there are still a number of examples of best practices that could enable airlines to improve their offers within the constraints they operate.

- Chapter 6, *Viewing the Changing World Map*, points out that there are at least a dozen emerging markets beyond BRIC that represent potential opportunities for airlines. If one assumes that the 19th century belonged to Europe and the 20th century belonged to North America, the 21st century is likely to be a global century with high-growth countries located all over the world. In fact, even the term BRIC has been expanded to BRIICS to include Indonesia and South Africa. This chapter discusses briefly the future of two regions, Africa and Latin America, as well as four individual countries, Chile, Egypt, Malaysia, and Turkey. The high-growth, and increasingly-competitive countries provide tremendous opportunities for not only airlines based in those countries, but also for other airlines to connect with these countries.

- Chapter 7, *Flying with Tailwinds against Headwinds*, deals with two categories of facilitators for succeeding in the new normal. The first facilitator is game-changing technology that can (a) increase the efficacy of airline operations, and (b) improve the customer experience of air travel. Examples of such transforming technology include new airplanes (such as the Boeing 787, the Airbus A350, and the Bombardier CSeries), smart mobile phones, on-board Internet capability, self-service facilities, and radio-frequency identification (RFID). The second facilitator is game-changing leadership, leadership that can make sense of changing trends and create value in times of chaos. It is acknowledged that airlines have been constrained in the area of business agility and the deployment of best business practices in light of their industry-specific constraints. However, the existence of inherent constraints simply means that potential winners must work even harder to find even more creative ways to work around these constraints. If the industry has been problematic since its beginning, surely renovation of ageing business models and strategies are needed to provide value for shareholders, not just for customers, labor, and other members in the air travel value chain.

The main audience of this book continues to be senior-level practitioners of different generations of airlines worldwide, as well as related businesses. It would also be of interest to those management levels aspiring to find unconventional ways to succeed. As with the earlier books, the approach is not based on theoretical concepts. Rather, it continues to be based on pragmatic analysis that is grounded, in turn, on the realities of the marketplace, and includes a collection of real-life case studies from different industries. As in the previous book, *Flying Ahead of the Airplane*, this book also attempts to weave together different topics and best global business practices and capabilities (synthesized from the experience of aspiring leaders) in a strategic way to provide, in a readable and down-to-earth book, value-adding insights for practitioners in the global airline industry. As in the previous book, this book also contains a large number of case studies, some in significant detail. Obviously, readers can skip the case studies if they are familiar with the businesses, without losing the main thought process. Furthermore, as in the previous book, this book contains a collection of forewords to provide added value for global readers through the different perspectives of knowledgeable and insightful executives in airlines and airline-related businesses.

Globalization will continue to uphold the need for air travel and the airline industry will definitely continue to grow. Management simply needs to learn to make money, at a sufficient level, and through the economic cycles. It can be done, and some airlines will make it happen by embracing the wisdom of the Chinese proverb: "When the winds of change blow, some build walls, others windmills."

Notes

1 The first, *Driving Airline Business Strategies through Emerging Technology* showed that in the rapidly evolving airline industry, emerging technologies could indeed play an increasingly critical role in the delivery of real and perceived customer value. The second, *Airline Survival Kit: Breaking Out of the Zero Profit Game,* wrestled with the precipitous decline in the profitability of the industry and discussed some strategies for dealing with the heavy burden of excessive complexity incorporated within the operations of legacy airlines. Having realized that the industry and the environment were experiencing step changes, the third, *Simpli-Flying: Optimizing the Airline Business Model,* drilled deeper into the discussion on restructuring of markets and the critical need for strategies to adjust to the new aviation realities. The central theme of the fourth, *Fasten Your Seatbelt: The Passenger is Flying the Plane* was that core customers—not airline management—are beginning to seize control of the direction of the industry. The fifth book, *Flying Ahead of the Airplane,* analyzed global trends and provided some thought-provoking scenarios to help airline executives adjust and adapt to a chaotic world.

Foreword

William Ayer
Chairman and Chief Executive Officer
Alaska Air Group

Americans have a love-hate relationship with our industry. News media coverage is typically negative and stand-up comedians reduce airlines to a punch line. But the fact is commercial aviation represents a key driver of the economy and helps enable our lifestyles. U.S. carriers fly 2 million travelers a day and directly and indirectly generate some 10 million jobs. The industry has delivered unprecedented safety and reliability with frequent and convenient schedules, and, through competition, has offered fares that have made flying affordable for virtually everyone.

Unfortunately, the economics have not been favorable for most airlines. Today's industry is challenged like never before and its future is ever more uncertain. The period following 9/11 brought multiple bankruptcies, and the oil price spike of 2008 and subsequent collapse in demand during the worst recession in decades has threatened even the healthiest carriers. Our economy will likely recover—slowly—but the fate of the airline industry depends on what its leaders and all other stakeholders do now.

Make no mistake that change is required—not just incremental improvement, but wholesale transformation—to satisfy the needs of employees, customers and investors. Some changes are underway, but much more remains to be done. The responsibility rests with the industry's leadership and it starts with long-term goals. Of course, it's difficult to think long term when short-term survival is in question. So liquidity is the first, urgent order of business. Cash is king.

Long-term progress is best achieved by focusing on what's controllable. It's tempting to dwell on dismal news, look to others for solutions, or just hunker down and hope things get better. Great companies don't use hope as a strategy. Instead, they take advantage of difficult times to improve.

The goal of transformation is to provide long-term benefits for all stakeholders: secure jobs and retirements for employees, excellent value to customers, and—for the first time in this industry—a consistent and reasonable return to capital providers. As alignment increases, the transformation benefits from a virtuous cycle among these groups. Other stakeholders—including governments, airports and labor unions—also need to make changes. As they do, they'll realize greater alignment, too, and the benefits of "win-win" partnerships.

In certain remote areas, such as the state of Alaska (and parts of Africa and Latin America discussed in Chapter 6 of this book), air service is the only means of transportation and represents a social and economic necessity. Airplanes fly everything from medical patients to high school basketball teams, mail, food and much more to provide these residents with a vital link to the outside world. That neighbor-helping-neighbor environment, where adversity is a daily fact of life, has helped our company evolve and learn many lessons for successfully navigating today's turbulent skies.

Alaska Air Group is fortunate in having terrific employees who are willing to try new things to re-engineer our business. We were the first airline company to sell tickets over the Web and offer widespread use of kiosks for airport check-in. We also pioneered GPS flight deck navigation to enhance safety and reliability. Unlike most of our peers, we avoided bankruptcy following 9/11 and are gaining important capabilities by working with our people to make needed changes. We're flying a new fleet of fuel-efficient Boeing 737s, have restructured our network and are serving many new markets, including Hawaii.

We also recognize that customers ultimately determine our success, so we're striving to be closer to them and their communities and to adapt faster to changing needs. While we are realistic about the challenges still ahead, we're more optimistic than ever because of the dramatic improvements we've made over the last decade.

This world needs a strong airline industry. But the changes required, while very achievable, are not for the faint of heart. Nawal Taneja's new book provides the encouragement to take bold steps. The payoff will be a transformed industry that's in sync with today's needs.

Seattle, USA
November 2009

Foreword

Jim Compton
Executive Vice President and Chief Marketing Officer
Continental Airlines

I know at Continental we really would have very little to market and very little to sell if it were not for the focus that our colleagues in operations bring to the airline each and every day. It is the experience that we market. Clean, safe and reliable are the cornerstones of our company that are also the very elements that allow us to be successful at Continental. However, the constantly changing marketplace has made the goal of reaching sustainable profits fleeting. Dr. Taneja once again in this book successfully discusses and provides ideas on how managers should think about that changing marketplace.

How nice it would be to have sustainable profits by adding up our costs and then add a little more to re-invest in our product and our people. Based on the industry's performance over the past 30 years, employees and shareholders would be much better off. The reality is we are no longer a cost based industry. We are now a market based industry. The marketplace sets the prices we charge and the simple rules of supply and demand dictate the performance of the industry.

One of the many reasons Continental has been successful over the past 15 years comes directly from our annual *Go Forward Plan*, specifically our *Working Together* cornerstone. At Continental providing direct, frank and honest communication throughout the company is the foundation of our success. All of us at Continental understand what we are trying to achieve and how we are going to go about achieving it. Our plan each year is very "transparent." Bill Brunger, a former colleague at Continental has said that the internet has been one of the significant changes to the industry in the past 10 years. It has provided customers full price transparency that previously was more difficult to find. Pricing search engines are so powerful today that the best price options are on display on consumer's computer desktops and PDA's.

Marketing can learn much from those two themes. Customers want direct and honest communication and technology has allowed us to use the aforementioned transparency to offer choices to our customers. Air travel is not a commodity. It is an experience. How we present our product, the choices we put around our product, the communications to customers as they travel and use our product, all define the experience. The higher the experience the higher will be the consumers' willingness to pay premium prices. The fundamentals of the airline revenue generating systems have not changed. We are still an industry that is dependent on segmentation. The customer is choosing a level of experience. eCustomers are also demanding that we deliver on the experience that they pay for and that they expect.

The industry as a whole needs to achieve sustainable profits. The need to drive costs down through efficiencies will continue, but the successful airlines will be the ones who on top of cost savings, through efficiency, drive a revenue premium by developing a positive experience that customers associate with an airline's brand. At Continental we realize that this is a challenge every day. Moreover, the challenge is growing and evolving as consumers increase their use of the Information and Communication Technologies. Dr. Taneja's book lays out many aspects of this changing marketplace where consumers are demanding more and more.

Houston, USA
January 2010

Foreword

Enrique Cueto
Chief Executive Officer
LAN Airlines

Aviation in Latin America has come a long way. Twenty years ago, the industry was mostly dominated by state-owned flag carriers with inefficient operations that provided passengers with a product that left much to be desired in terms of reliability, punctuality and service. Today the scenario is very different. Various leading airlines have emerged within the region, with successful business models that have earned international recognition. Four Latin American carriers are today publicly traded on the New York Stock Exchange. How did this change come about? The answer, as Professor Taneja clearly explains in this book, is *innovation*.

Two decades ago, LAN Airlines (historically known as Lan Chile) was a small airline operating out of a country with less than 13 million inhabitants, inconveniently located at the southwestern corner of South America. Based on these demographics and on the traditional airline business models, nobody expected LAN to become the profitable and efficient airline it is today, among the world's top ten in terms of market capitalization. As Professor Taneja candidly states, aviation in Latin America faces various challenges, including the small size of its markets, ownership restrictions and inefficient infrastructure. Most countries in the region (except Brazil and Mexico) each represent only close to 1 percent of world traffic. However, Latin America is also one of the fastest growing regions in the world for air travel, with significant growth potential due to its low enplanements and growing levels of GDP per capita. This growth is further enhanced by the fact that Chile, Peru and other countries in the region are increasingly adopting solid macroeconomic policies that support continued growth and stability and thus stimulate a growing demand for air travel.

The LAN business model has sought to contribute to the development of aviation in the region, providing the best connectivity and quality of service for Latin American passengers. In order to do this, together with local investors we have expanded to other countries using a common corporate image under the LAN brand. In this way, we have been able to generate the necessary economies of scale to support a successful and efficient operation, which has ensured the ability to finance significant fleet expansion and renewal in order to maintain a modern and fuel efficient fleet.

The first step in LAN's regional expansion was the launch of LAN Peru in 1999. This was followed by LAN Ecuador in 2003 and LAN Argentina in 2005. Each of these airlines has domestic passenger operations in their home countries, as well as international operations to various destinations within Latin America, the United States and Europe. Similarly, LAN's cargo operation is based in Miami and has affiliates in Brazil, Mexico and Colombia, creating an integrated network with broad coverage throughout Latin America. The approach LAN takes to integrating its passenger business with the transportation of air cargo is another distinctive element of the Company's business model. The belly space of all long haul passenger aircraft is filled with cargo, maximizing the utilization of the aircraft and increasing the efficiency of our operations by reducing the breakeven load factor. This is a competitive advantage of LAN, especially on long haul routes where cargo operations are more significant. As a result, cargo has historically represented approximately one third of LAN's consolidated revenues, a much higher percentage than the industry average.

In 2007, LAN Airlines and its affiliates introduced a further innovation by implementing low-cost type practices in their respective domestic operations. This implied a revolutionary change, defying well established practices and perceptions that had been unquestioned by the Company for years. As the LAN carriers renewed their domestic fleet and completed the replacement of the Boeing 737-200s, aircraft utilization increased. The sales and distribution process became focused increasingly on on-line sales through LAN's website, and the on-board service process was expedited, replacing hot meals with a cold snack, which in addition allowed for the replacement of the galleys in the aircraft with two more rows of seats, thus further reducing unit costs. All these efficiencies were passed on to passengers through lower fares, which have led to an explosion in terms of traffic growth, with two consecutive years of approximately 20 percent growth in the domestic markets of Chile and Peru for LAN Airlines and LAN Peru, respectively. This was a dramatic change not only for the airlines, but for the industry as a whole, contributing significantly to the development of air travel within the region.

Airline business models are constantly tested as the industry is affected by the many external shocks that in one way or another impact demand for passenger and cargo transportation. Over the past 15 years, due to its flexibility and constant innovation, LAN has proven resilient to the various crises that have affected the industry – including recession, terrorism and rising oil prices – and is one of the few airlines that have shown consistent profitability. The year 2009 has been one of the most challenging in the Company's history as we faced a difficult environment in all our markets, including a global recession that affected both the cargo and passenger businesses, as well as the AH1N1 epidemic. However, these multiple shocks find us on stable footing based on the groundwork we have constructed over the past years. As such, although revenues have dropped, we continue to show solid profitability and double digit operating margins, a testament to the resiliency of our geographical and business diversification, efficient cost structure and flexible operations.

Despite the advances and significant development of the Latin American aviation industry, we still face challenges ahead. There are steps that can be taken throughout the region in favor of liberalizing regulatory controls as well as continuing to improve and develop airports and related infrastructure. Joint efforts towards these common goals will ensure the continued development of the aviation industry and will support the economic development of the region as a whole.

Santiago, Chile
December 2009

Foreword

Hugh Dunleavy
Executive Vice President, Strategy & Planning
WestJet Airlines

The air transportation industry, as it has existed for the past 30 years in Western Europe, North America and much of the Pacific Rim, is broken. Too many airlines, with too many aircraft, and not enough infrastructure are chasing too few passengers who only want to pay too little money. This coupled with a supply chain in which almost all of the links are monopolistic suppliers, each taking their own pound of flesh from the industry, and a degree of governmental regulation that may only be rivalled by the nuclear industry, airlines today are facing unprecedented challenges.

Over the past couple of decades, when airlines got into financial trouble, they would often go into bankruptcy protection in order to restructure their balance sheets and trim their costs. Despite the recurring theme of airline troubles, it was very unusual for an airline seeking recapitalization to have much difficulty in securing new sources of funds. The world financial meltdown of 2008 and the resulting seizing of the capital markets have apparently closed off access to recapitalization funds. Had the cost of oil not plummeted from its 2008 high of $147/barrel, it is likely that we would have already seen one or two major airlines fold operations.

The classical model for airline revenue generation has, both figuratively and literally, split demand into two major classes: the high fare front of the cabin and the low fare back of the cabin. The bulk of the passengers in the back of the cabin was charged fares that were clustered around the break-even mark. Achieving a decent load factor in the back of the cabin essentially allowed the airline to recover its operating costs. It was the front of the cabin, either business or first class, that enabled the airline to generate a profit. With the collapse in demand that occurred in the later part of 2008 and into 2009, not only were airlines unable to hold the prices for the back of the cabin passengers (while still achieving decent loads), but the demand for the front of the cabin suffered even greater losses despite a substantial drop in fares charged to the business traveller. Some estimates for 2009 business traffic are suggesting that this segment of the market may be down in the 15–20 percent range for the year and that this traffic may not recover until 2011/2012, as discussed in Chapter 1 of this book.

Since the last major crisis in the airline industry following 9/11, airlines have made significant progress in reducing their operating costs. Many of the major carriers have gone through at least one round of bankruptcy protection and reorganization and, as a result, have managed to reduce their labor costs, rationalize their fleet, restructure their remaining aircraft lease/purchase obligations, and reduce their fuel consumption. In fact, it could be argued that there is little room for further improvements in most carriers' costs of operations. Given all this, it is reasonable to question whether or not it is within the power of any individual carrier to salvage a reasonable future. Perhaps the future lies with the industry as a whole coming up with a new model that recognizes the following:

- suppliers to the industry cannot earn profits while the industry itself suffers continual losses;

- that the various governments and regulatory bodies cannot continue to apply more, and more complicated regulations, fees and taxes on an overburdened industry;

- that passengers must expect to pay the cost of travel;

- airlines start measuring success based on profitability instead of market share.

A revenue generating approach that appears to be gaining ground in the airline industry is known as "a la carte" pricing. This model operates by unbundling all service offerings (that were historically included in the price of the ticket) from the base price of a ticket and offering these services as additional options that the consumer can elect to purchase. The years 2008 and 2009 saw a significant expansion in the adoption rate of this model, as a large number of the world's airlines sought to significantly increase their revenue stream through the introduction of more and more service fees. This was further compounded by the extremely quick run-up in the price of oil and the associated refining costs for jet fuel which triggered significant fuel surcharge fees on most tickets.

When oil prices collapsed in 2009 there was, as expected, no rush by the legacy airlines to remove either those ancillary fees or the large fuel surcharges. This classical approach by legacy airlines to antagonize their customers appears to continue unabated. It should be noted that during this period of high oil prices, a number of airlines, including some legacy airlines, attempted to hold the line and not charge these fees on the basis that it allowed those airlines to further differentiate themselves from the herd and allowed them to deliver an enhanced value product for the consumer. The fundamental assumption underlying this approach was that the consumer, as a rational economist, would naturally evaluate the price of the ticket plus all the ancillary fees and compare total costs across the competing airlines and then choose the lowest total cost solution that met the individual's travel requirements.

Daniel Kahneman's Nobel prize in Economics on "Prospect Theory" in 2002 demonstrated the fact that consumers are not rational economists and that even if they get their complex sums correct, they do not act as if they do. This was amply demonstrated by the number of airlines that ultimately implemented additional ancillary fees simply because the consumer was not differentiating between an airline that charged fees and an airline that did not charge fees.

A more fundamental problem for the legacy carriers is that they have not kept up with the pace of change in the industry and, in my opinion, have had their collective heads in the sand for many years, in their belief that they could isolate the impact of the LCC's into a relatively small segment of their overall business, and, for those markets not having an LCC competitor it was basically business as usual. This mindset at the legacy airlines must change. The LCC business model is continuing to evolve and recent announcements by airlines such as Southwest and WestJet on a codeshare partnership demonstrated the commitment by the LCC's to provide service to a wider customer segment and to offer a competing network that dramatically expands both the scale and scope of their operations. I believe we have only seen the tip of the iceberg in terms of the evolution of the LCC business model. Take a look at Figure 7.4 in this book that illustrates the "Potential Opportunities for Transatlantic Service by Ryanair."

In addition, in a recessionary period the LCC's have traditionally faired much better in economic terms than the legacy airlines, and, in fact, have used these downturns as staging points for the next spurt in the LCC growth cycle. The years 2010 and 2011 will be telling years for the survivability of the legacy model, their ability to respond to the LCC's and the ultimate structure of the LCC model.

One of the challenges for the legacy airlines is the fact that many of the LCC's, equipped with modern fuel efficient fleets and a significant cost advantage over the legacy airlines, were able to avoid implementing many of the these service fees and provide a superior quality product and service compared with the legacy carriers. Have the legacy carriers simply gone too far in their efforts to reduce costs that they now have become the poor relation compared with the LCC's? At WestJet we believe that the quality of our products and services are what differentiates us from our competitors and that we offer service levels considerably superior to the legacy airlines in North America.

In my opinion the LCC model is continuing to evolve and the more forward-thinking airlines are investing heavily in technology, products and services to create a new hybrid model that can address all the needs of the savvy business traveler and can offer a virtual route network through airline partnerships that can rapidly develop into a true competitor to the legacy-based airline alliances. These partnerships will ultimately deliver a global network capability at much lower costs than that possible by the legacy airlines.

Airlines such as WestJet have already announced interline deals with major international airlines that allow network flow from Europe, Asia and the United States into and out of Canada. The continued growth in the LCC airlines from their traditional domestic markets represents an increasing and significant threat to the long-term viability of the legacy airlines. As noted earlier, it is not clear just how much more costs the legacy airlines can extract from their business models and simultaneously obtain improved flexibility in work rules to allow them to become competitive with the increasingly aggressive LCC's.

This leads naturally into the need for the world's airlines to enhance their existing revenue management tools. The classical airline RM tools are in need of upgrading as more and more markets are subject to non-segmented pricing structures, and the ability of the traditional forecasting and optimization approaches are unable to adapt and respond quickly enough to the prevailing competitive pricing conditions.

As discussed in Nawal's book, the world economic environment is undergoing dramatic changes, new governance rules are likely to be imposed on financial institutions, further expansion of Open Skies agreements and removal of the old limits on foreign ownership will be major incentives for a greater degree of cross-border mergers that will further cloud the waters of industry change.

Further, an ever looming challenge for the entire aviation industry is the environmental file with various entities seeking a leadership role in setting appropriate emission targets and taxation structures which will further dampen the growth in air travel.

As the Chinese proverb states: may we live in interesting times, could not be more true than in the airline industry space.

Calgary, Canada
December 2009

Foreword

Peter Harbison
Executive Chairman
Centre for Asia Pacific Aviation

Defining long term aviation trends is always heavily coloured by the currents of thought which fill the present. Over the past couple of decades I have been involved in several such exercises and each time there has been a serious event which has filled the short term horizon.

The deep recession of the early 1990s, the Gulf War, the Asian Financial meltdown, the tech bubble burst, September 11, SARS and now the GFC, the global financial crisis. Aviation, probably more than any other industry, is enormously susceptible to crises, be they economic, medical, terrorist-sponsored, or simply internal revolutions such as the impact of low cost airlines.

A few years ago I named this the "constant shock syndrome"; because of it, aviation rarely seems to have an extended period where it actually grows at the so-called long term growth rate, instead constantly slumping and rebounding.

So, when looking forward, it is important to avoid falling victim to assumptions that the existing crisis will continue to influence events in five years time. We may be "losing" one or even two years of growth in today's deadly economic climate, but history suggests the reduced rate is unlikely to persist.

All of this makes life very hard for airlines and for airports – and their suppliers - each of whom must make major long term decisions about capital expenditure. Putting the immediate-term colouring of the current shocks in their correct perspective is always a severe challenge. So, in the darkest days of SARS, when some major Asian airlines were literally days away from grounding their entire fleets, making an assumption that it was only a temporary phenomenon seemed a brave call. Yet, a mere three months later growth had rebounded to the redoubtable long term trend.

Today the confronting issues are the "death of premium traffic" and the erosion of profitability of network airlines by low cost competition, as liberalisation spreads across the world. The combined result is an even more pressing need to adapt an airline model – and regulatory structure – that is patently unable to provide adequate financial returns. (But it seems that such a seismic shift is unlikely, even in the drastic conditions of 2009. At least two major flag carriers are being bailed out by their governments, who found themselves confronted with the unpalatable alternative of a major airline collapse.)

Meanwhile however, these setbacks have only temporarily stolen the limelight from one most indelible life-changing shock: the need to adjust to the priorities of climate change.

Whether or not premium traffic does return and, later, whether the basic low cost model can survive a world where fuel prices above USD150 greatly dilute the value of that model, there is one thing certain. The climate imperative will force our medium term projections at least to be adjusted downwards. If nothing else, the mounting levels of indiscriminate "environmental" taxation on the airline industry will force fares upwards; at worst, popular pressures will reduce leisure travel, despite the massive value of tourism, the world's largest industry, to developing nations in particular.

Beyond the medium term, around 2020, assuming that alternative fuel sources and other ways of reducing harmful emissions can be developed, the curve should tick upwards again. But I suspect those innovations will take at least a full decade to implement. So we may, for once in fact be confronted by an awkward decade which does not conform to history.

But there is another major shift in the world equilibrium. This one is geographic. Nawal discusses this in Chapter 6 entitled, "Viewing the Changing World Map." Asia is in the ascendancy, economically and in many other ways. This is nothing new; for 18 of the past 20 centuries, China has been the largest economy in the world. Today it is well on the road to reclaiming that mantle.

And China, with its 1.4 billion people, is only one of a number of emerging Asian nations, including Vietnam, with a 100 million population, Indonesia, with 240 million people, not to mention India's 1.2 billion, all of whom are fast developing to a stage where substantial proportions of their populations have the ability and the desire to travel.

The scope of this potential growth defies forecasting. There is simply no basis for projecting just how big these countries' air travel markets will become in the next couple of decades. Given favourable conditions, the rate of expansion will be without precedent. Moreover, just as economic expansion will push growth, so the relaxation of regulatory controls on air services will fuel the flame, notably in and between the wealthy northeast Asian markets.

One indicator of what might be to come is the fact that, despite a decade of annual high double digit passenger traffic growth in the domestic Chinese market, some two thirds of that air travel is still business related. The leisure market has barely been tapped. The rapidly rising average per capita GDP of Beijing's 17 million population and Shanghai's 19 million is already well above the threshold of air travel profiles. The country boasts at least 100 cities with populations exceeding one million people (the China Daily says 700); by contrast the US has 10 cities of this size.

Add in the potential growth of the southeast Asian countries, India and the Middle East (which between them account for the vast bulk of new aircraft orders) and it is clear that the old order, of European and American leadership, will be no more. See, for example, Figure 7.5 in Nawal's book.

In these two main ways, over the course of the next decade the aviation world will undergo upheaval far greater than ever before. Nawal describes this as the "New Normal."

As for the demise of premium travel, in markets like Asia and the Middle East, where most future aviation growth will be, I would suggest that this is one reported death that has been greatly exaggerated. There are literally millions of millionaires who will prefer not to travel in the back of the aircraft – unless it is their own private jet.

Sydney, Australia
November 2009

Foreword

Henry H. Harteveldt
Vice President, Principal Analyst
Forrester Research, Inc.

The global airline industry seems to be perennially at cross-roads. We have careened from one challenge to another, from deregulation to alliances to changing regulatory environments to terrorism to the environment to technology—with economic booms and busts constantly scattered throughout.

Airlines are now at yet another cross-road. Amidst a global economic crisis that has affected business and leisure travel, premium and discount, airlines must contend with a very different customer—one who is increasingly digital, fickle, and demanding. When consumers can choose from tens of thousands of possible beverage permutations at their local Starbucks, access more than 100,000 applications to download to their iPhone (which itself comes with different storage capacities and colors), or configure their iGoogle page with color, themes, and content arranged precisely to their interest, industries that aren't focused on delivering tailored customer experiences fade to the background.

Airlines need to recognize that they are increasingly serving an experiential traveler—people who appreciate convenience, control, and creativity. Data from Forrester Research's North American Technographics® Travel Online Survey, Q1 2009 shows that 45 percent of US leisure airline passengers are primarily motivated in their lives by entertainment. Don't misinterpret this as meaning your customer values only in-flight entertainment. It means you need to build your product, pricing, technology, and communications to successfully serve a customer who views the world and everything she or he buys through a lens framed by interactivity, control, constant communication, and the expectation—reasonable or not—that what they want, they *will* get, whether from you or a competitor.

Nawal has aptly titled this book "*Looking Beyond the Runway*." Airline executives and managers must look beyond their operations (and spreadsheets), and strive to create strategies in which the customer considers herself or himself to be in a partnership with an airline (or alliance). Recognizing and finding ways to serve the experiential traveler consistently across all touch points such as Web sites, mobile, loyalty programs, in-flight—even something as utilitarian as baggage reclaim—will help smart airlines connect with passengers and contribute to a change in how they view airlines. Airlines that understand and fulfill experiential travelers' key rational and emotional needs stand to win by shifting that customer's viewpoint of airlines from a "back of the mind commodity" to something they focus on more actively. Make your airline one that experiential travelers desire, and you reduce the importance price plays in the purchase decision, while hopefully benefitting from greater loyalty and more revenue.

San Francisco, USA
November, 2009

Foreword

Pedro Heilbron
Executive President and Chief Executive Officer
Copa Airlines

The airline industry, which has historically been brutally competitive, now finds itself at yet another critical moment. Over just a little more than a year, airlines have confronted not only each other in spirited competition, but also record high fuel prices, an economic crisis that resulted in plummeting demand and a flu pandemic.

It is clear that to survive these challenging times airlines must adapt to a "new normal".

This new normal is forcing airlines around the globe to adapt to these ever changing challenges at lightning speed. As in any industry, the customer is the boss; and our customers have come to expect more and better service at ever lower prices. Airlines have continued to evolve to meet these expectations, but as expectations are met they only continue to rise.

Low cost carriers have forced "legacy carriers" to do more at a lower cost – this has been a positive for all parties. We are now a more efficient and customer focused global industry. I and all of my co-workers at Copa Airlines enjoy these challenges which force us to wake up every day and think how we can do more with less.

As carriers search for much needed revenues to survive, they can never lose sight that customers must see value in what they pay for. Low cost carriers are clear in their value proposition and as such customers reward them with loyalty. Legacy carriers are now repositioning themselves to help adapt to this new normal and in some cases reinvent who they are.

Copa Airlines has always been clear in our value proposition – our goal is to provide safe, reliable and customer friendly air transportation within the Americas. Our customers appreciate this simple focus. As we prepare for the "new normal" we at Copa

Airlines believe in many ways certain truths still prevail and will remain the same: you must consistently provide good service at a fair price. This approach is not, however, at the expense of evolution. Copa continues to look, not only within our industry and region, but around the globe in every industry to benchmark companies that excel in superior customer focus and cost control. As a small Latin American airline surrounded by larger global carriers, our ability to compete relies on execution and clarity of purpose. Adapting to evolving customer needs has been and will continue to be our first priority.

Dr. Taneja is a thought provoking expert who greatly helped Copa rethink how we approach our business. His ability to challenge commonly held beliefs and force progressive thought has been key to Copa's strategies and evolution. From all my co-workers we wish to commend him for his continued efforts to make airlines everywhere better equipped to exceed customer expectations and continue to be the lifeblood of cultural and commercial interchange around the globe.

Panama City, Panama
November 2009

Foreword

James Hogan
Chief Executive Officer
Etihad Airways

During nearly three decades in the airline industry, it would be hard to recall a time when work was ever just business as usual. The industry's first turbulent century has felt keenly the impact of wars, recession, oil price volatility, terrorist attacks, extreme weather, and more recently, the effects of a global pandemic.

In this dynamic and ever-changing environment, it is optimistic and maybe even foolish to take 'normal', or what is now called the 'new normal', for granted. The constant challenge for airline management is how to navigate this shifting, inconstant world.

New wave carriers undoubtedly have the upper hand in this very fast changing environment. Without the constraints and history of older legacy carriers, younger airlines have been able to use 'best practice' as their baseline in the creation of modern relevant business models that are better geared for the global economy. The importance and implementation of best practice, whether in customer experience or innovation is strongly endorsed throughout this text and illustrated in case studies and numerous examples.

I have been fortunate in the last seven years to play a role in the growth of a new wave carrier and the newly emerging Middle Eastern aviation industry. In the Gulf region, the fortuitous combination of good geography, new technology and bold investment underlies the rapid and successful rise of the modern aviation super-hubs that today bridge the globe.

Middle Eastern carriers have joined the top ranks of the airline industry and they're here to stay. Serving not only a region where prodigious economic development continues to create demand for travel, Middle Eastern carriers have earned a reputation for innovation and service excellence in the international market where they are forcing a rethink on traditional airline models.

For example, Etihad Airways uses benchmarking outside the industry to good effect in its product and service design, and today the brand, and the underlying inspired service concept, finds its inspiration in the wider world of hospitality – in five star hotels and fine dining – to create sustainable competitive advantage.

Innovation has to be accompanied by an unerring focus on the customer and a corresponding commitment to deliver great customer service in all cabins consistently.

Even in the most challenging of times, cost should never be an excuse for compromised product and service. Today long-term strategy dictates that measured investment in fleet and in-flight product continue even in the lean times, as much for our customers, as for our shareholders.

Despite the industry's considerable success in cutting costs since the catalytic events of 9/11, there is always more to be achieved. To address the commercial imperatives of a perpetually unprofitable industry, the business will need to seriously evaluate and consider consolidation. Not necessarily in line with the European models, but certainly to the extent that we make better use of synergies and economies of scale to drive down cost.

Although individual brand success and competitive advantage are important indicators of success, our survival as an industry in these changeable and challenging times will depend increasingly on our ability to collaborate through shared knowledge and resources and collective thinking.

Professor Taneja's comprehensive analysis of thinking from within and outside the industry provides this collective wisdom of ideas and practice, and is therefore essential reading for airline management facing the 'new normal'.

Abu Dhabi, United Arab Emirates
November 2009

Foreword

David V. Jones
President and Chief Executive Officer
Amadeus IT Group

As the airline industry scrutinises the global economy for signs of recovery Nawal Taneja's lessons from other industries are illuminating and thought-provoking.

I would agree with Professor Taneja that "enabling technology capability" is one of the, "three fundamental growth drivers in commercial aviation". But, since the book looks across the airline industry, technology in this context is as much about increasing aircraft efficiency as it is about web 2.0 and mobile technology.

I would like to shine a light on the specific question of IT infrastructure: reservation, inventory management and departure control systems. Whether you call this trio of applications a Passenger Service System or, as we do at Amadeus, a Customer Management Solution (CMS), it is absolutely business critical. When British Airways upgraded to our Altéa CMS, Paul Coby, BA's CIO, called it "open-heart surgery for an airline".

Given the importance of such systems, it is no surprise that airlines proceed with caution when it comes to upgrading them. But upgrade them they must. As more and more bookings go online, so the ratio of transactions – operations performed by an airline's reservation system – to bookings increases enormously. Here are some figures from our system. In 1999 we processed on average 30 transactions for every booking made in our GDS. Today, we process 160 transactions for every booking. So our systems need to be five-to-six times more powerful today than they were ten years ago to handle the *same number of bookings.*

Airlines running their own IT systems face exactly the same problem. But it is well-known that many of today's Passenger Service Systems are based on technology designed in the 1970s and 1980s; they are not prepared for today's environment, let alone that of the future.

Airlines are faced with a number of choices at this stage. Do they make significant capital investments in improving their existing systems? Is there value in retaining their in-house systems? Do they look at outsourcing to provide increased flexibility while removing the need for major capital expenditure?

Professor Taneja points to the potential benefits of Cloud Computing as it enables, "an airline's IT department to access applications and services, reduce an airline's IT department's capital costs, and enable it to pay on a per-usage basis."

Amadeus' approach to this concept – which, incidentally, goes back to our original formation in 1987 – is to outsource customer management processes to a common platform, which is a single system and repository of information shared by a number of carriers. Some commentators suggested that outsourcing to a common platform both increases risk and reduces competitive advantage if all customer management processes are shared.

These concerns miss the point. It is commercial agility and insights that deliver competitive advantage in today's market environment. The concept of a common platform for customer management systems, which represents the biggest shift in airline IT in twenty years, does not reduce competitive advantage. Instead, ensuring that customer management systems are hosted on reliable, functionally rich, next generation systems means that one of the most significant IT burdens is removed and that airlines can focus on what the customer wants and what will deliver real competitive advantage whether that is the introduction of new services, routes or partnerships.

With the new generation of Customer Management Systems, competitive differentiation is enhanced through technological or commercial collaboration. The scope for differentiation exists because airlines can determine how they choose to use the tools they share. Indeed, as airlines work together more closely, either through mergers and acquisitions or through strategic alliances, the need for an IT infrastructure which enables commercial collaboration becomes ever more acute.

In fairness, many of the world's airlines have now adopted this community approach to IT infrastructure. At the last count 126 airlines carrying nearly 550 million passengers a year use Amadeus' community-based reservations system, of which

63 airlines, carrying over 270 passengers a year, also use our community-based inventory management system. These airlines have all come to the conclusion – albeit with a little help from my sales team – that their commercial interests are best-served by pooling development costs on a single IT platform.

I defy readers of this book to find another industry which has challenged the conventional wisdom on competitive strategy to the same extent, and come to the conclusion that it makes commercial sense to share a common IT platform for mission-critical systems with the competition. Perhaps that is one positive lesson that the airline sector has for scholars of other industries.

Madrid, Spain
November 2009

Foreword

Bob Jordan
Executive Vice President – Strategy and Planning
Southwest Airlines

No word better describes the environment faced by airlines this decade than *instability*. Since 2000 the industry has faced a wave of challenges: the September 11th terrorist attacks, rising and volatile oil prices, economic upheaval, rising costs, and changing consumer demands. Looking forward (be it a "double dip" recession, permanently high oil prices, or "cap and tax"—to name just a few examples) the challenging environment shows no sign of abating. That said, airlines have become increasingly adaptive over the past decade, a significant departure for an industry where substantive change was the equivalent of "turning the battleship" so to speak. The increased use of self-service to both decrease costs and increase Customer choice, new product and service offerings, new pricing techniques, and increasing scheduling volatility are just part of the "New Normal" for the industry this past decade, as described in Dr. Taneja's timely book.

Despite our success, Southwest Airlines is not immune to that environment. While beneficial fuel hedge positions helped cushion the blow for a while, reacting to the new normal has driven a pace and degree of change at the company greater than any other time in my 22 years with the airline. Our Strategy since 2006 has been "Transform Southwest for a Changing World." Notice the key word is *transform*, not *change*. To *transform* is to leverage the tremendous strengths we have at Southwest Airlines—our safety record, our low-fare brand, our focus on excellent Customer service, our financial position, our unique Culture, and most importantly, our People – to build a stronger, more adaptive company. To *change* can often mean to pursue something new without being grounded in the essentials that provide your core strengths; i.e. simply change for change sake. At Southwest our approach is to rapidly adapt, while still maintaining and improving our core operational, financial, and cultural strengths. Improved boarding and new gate lounges, new products such as Business Select and Early Bird Checkin, an improved southwest.com website, a commitment to RNP GPS based navigation and satellite based inflight internet connectivity, leveraging new communication channels with our Customers and emerging social media, and numerous network techniques that both improve the marketability of our flight schedule and squeeze more product out of the same number of aircraft are all examples of transformation underway at Southwest. Doing all of that while also strengthening the core: our low-cost leadership, our productivity (Employees per aircraft is down from over 90 in 2002 to about 64 today), our excellent operational record, our financial strength, and our Culture is a testament to the People of Southwest Airlines.

That said, in this industry you can never rest. The challenges and pace are increasing, not decreasing; therefore the need to react effectively is greater than ever. The key is balancing longer term goals for delivering value against the flexibility to adapt in the short term. New products, processes, techniques, and technology can often take years to develop. Likewise, it can take years to fully understand the impact on the brand, demand, or Customer perception of decisions made in the short term.

A relatively recent industry example is the unbundling of the traditional product and a literal explosion of new fees as a way to drive revenue production. At most airlines, passengers face fees for everything from talking to a human being, to checking a bag, to a snack or soft drink in flight. Some fees, such as premium seat selection, are for new services; most, however, are simply a fee for a service that has been traditionally provided for free within the price of the ticket. Many argue that fees have no impact on demand; that is, Customers willingly pay the fees. History, however, particularly in the era of price transparency via the internet, would say that consumers are smarter; ultimately understanding the total price paid for their journey, and will adjust their travel and airline choice decisions based on that. While significant product unbundling and fees for service can add to the bottomline in the *short term,* the impact on the brand, and therefore demand and revenue production over the *long term*, could be significant. At Southwest we have chosen to very carefully balance the desire for new revenue sources against our desire to continue to build long term value and loyalty to our brand as a differentiator.

The bottomline? The need for and pace of innovation (as discussed in Chapter 4 of this book) will increase, not decrease, over time. Knowing your strengths and your challenges, having and crisply executing a long term plan based on that knowledge, and maintaining your ability in the short term to be flexible and react is key. The People of Southwest Airlines have a legacy of resilience and innovation that has made air travel fun and affordable for Americans. I have no doubt they will continue to rise to the challenges facing our company and our industry in the future.

Dallas, USA
November 2009

Foreword

Alan Joyce
Chief Executive Officer
Qantas Airways

For some time now, Nawal Taneja has been challenging aviation executives to go beyond the boundaries of our industry and seek out examples of global best practice across sectors as diverse as cars and luxury hotels, personal computing and fast food.

Airlines have been forced to confront the immediate challenges posed by the global financial crisis and subsequent economic downturn. In the case of the Qantas Group this has involved decisive management of our flying operations, capital expenditure pipeline, cash flow, and cost base.

But short-term imperatives must never be allowed to detract from long-term strategy. This is the reason why the Qantas Group has a rare advantage in having not one, but two, of the leading airlines in the world in their categories in Qantas and Jetstar. Since commencing in 2004, Jetstar has kept growing strongly and we have significant plans for Jetstar's future growth.

But we are also confident about the future of Qantas, our iconic, premium airline. Certainly demand for premium flying has been affected during the global downturn, and pretty seriously. But our research confirms that there is now, and will continue to be, an appetite for premium travel among both domestic and international travellers.

This does not mean that premium flying will return to business as usual. As Nawal Taneja reminds us, looking beyond the runway means finding both inspiration and practical precedent in the examples of other companies and industries. For example, the modern evolution of luxury has seen some of the world's top brands broaden their market reach without diluting their core values or brand equity. Success in such cases depends upon constant innovation supported by strong employee engagement, a close-up relationship with customers, and is underpinned by new and converging technologies.

For airlines to evolve as premium carriers, we must be ready to re-engineer how our products and services are delivered, even in radical ways. The next generation premium airline cannot win if it operates in the same way as many of the lumbering legacy businesses of the past (and some in the present).

Across Qantas we see opportunities to innovate in everything from our information technology processes to aircraft configuration to fuel conservation; from fleet simplification to supply chain management. Incremental change won't be enough: people, processes and technologies must all combine to deliver maximum punch, at minimum expense. Our loyalty program, Qantas Frequent Flyer, has been boosted by a series of innovations, including powerful commercial partnerships that enable customers to earn and redeem points everyday. We have an ambitious project underway to deliver radically improved airport processes to deliver speed and ease for our domestic premium customers.

Nawal Taneja challenges us all to look beyond old thinking and take hold of new insights and opportunities. These are tough times for global aviation, but there will be a bright future for those airlines with the energy and acumen to deliver the next generation of aviation excellence.

Sydney, Australia
December 2009

Foreword

Wolfgang Kurth
Member of the Board
Air Berlin Group
(Former Managing Director of dba, Hapag-Lloyd Flug Express, and Hapag-Lloyd Flug)

"Because I can look beyond the hills!" was Napoleon's answer when asked for the root cause of his success as a commander on the European battlefields. Anticipating what the future may bring to the aviation industry is Nawal Taneja's mission in his new book *"Looking Beyond the Runway."* He identifies the weaknesses in overall airline performance and how the airlines are losing their daily battle for customer satisfaction, and, in turn, higher yields and profitability. In addition, Nawal pinpoints good examples from outside the airline industry, and, by doing so, gives airlines a first glimpse of "beyond the hills."

Tour operators and charter airlines in Europe have failed to anticipate the changes in the booking pattern of leisure travellers. Individual travel dates and modular packages put together by the tourist in the Internet have eliminated the need for tour operators putting the charter airlines under tremendous pressure of overcapacity. At the same time the growing impact of LCCs upon our industries was widely ignored. Charter airlines, dominating the cross-border air traffic in Europe for more than two decades, lost the battle with the LCCs without even fighting, because under the "control" of the tour operators, they did not develop brands, or products, or distribution systems. Similarly, one could ask, why have traditional airlines lost market share to the LCCs? There are many reasons, but the root cause is they did not look *"beyond the runway*!"

It is the prime objective of all top executives to lead a company through the uncertainties of a global market to secure the future of the company. The emerging market provides numerous unknowns, particularly within the airline industry. Consequently, it is not good enough for top airline executives to just administer their jobs. It is their key responsibility to lead their organisation, including staff, stakeholders and processes, to what they identify to be beyond the horizon. This implies good knowledge of the "Change Potential" of their airline. It is a fundamental dilemma in the airline industry that even when information is pretty reliable, many changes cannot be implemented in a timely manner. It takes time to drastically change the course of an airline. A heavy balance sheet, an inflexible fleet, or inflexible labour contracts combined with rigid scope clauses, will not allow an airline to respond in a speedy manner to changes in the business environment.

While no one knows what the future will bring when changes finally take shape, wrong conclusions from the information available at the time have triggered the decision for change. The launch of the B757 back in 1979 is a good example of "not the best option." In light of a global oil shortage, fuel efficiency was the paramount design objective for this twin-engine, mid-size jet. However, before the first aircraft was delivered, the fuel price began to fall again, leaving Boeing with an aircraft design that was expensive to convert into metal. And airline customers were not willing to pay a premium price for a super efficient airplane, if that price could not be offset by the lower fuel consumption.

Some airline executives say that "six months is long term planning in the aviation industry." It should be a strong message that flexibility should be one of the key design criteria for an airline organisation. Moreover, knowing that the majority of airlines are suffering from more than 50 percent of the total costs being fixed costs, managements should amplify all efforts to convert fixed costs into variable costs. The need becomes even more critical if we assume that the speed of changes affecting the aviation industry is going to pick up significantly in the years to come, particularly relating to distribution and passenger booking behaviour. Airlines must, therefore, be on the lookout. When studying Nawal's *"Looking Beyond the Runway"* carefully, it is obvious the airlines do not necessarily "own" the customers. Companies, such as Amazon, eBay, Expedia, and Google know customers, including airline passengers, much better than most airlines! Isn't that a new challenge to the airlines?

Finally, airlines have to accept that they are no longer in a position to shape the market. If the market requires certain products and services, and the Googles of this digital world can meet the expectations of this new generation of airline passengers, airlines have to be prepared to meet the needs of the marketplace. Otherwise they will fail! Since these changes can be implemented in the digital environment very quickly without advance warning, the only answer for airlines is *flexibility*. To quote Napoleon again: "No plan survives the first attack of an enemy." A detailed five-year plan will not be the appropriate strategy to cope with changes which are still ahead of us.

Berlin, Germany
November 2009

Foreword

Andrew Lobbenberg
Airline Analyst, European Transport Equity Research
Royal Bank of Scotland

The airline industry is facing grave pressure having battled through spiking oil followed in short order by the credit crisis and an extreme economic downturn. As we look ahead, hoping without any clear-cut confidence for a recovery in demand, the industry looks in some ways to be improving.

Structurally we think that consolidation and the development of economically relevant joint ventures could offer a path towards an industry characterised by fewer independent players, greater rationality and the potential for stronger returns. But, whilst some of this improvement could come from cost synergies, much would come from revenue benefits, a fact not lost on the competition authorities, who are scrutinising all deals with vigour.

Operationally, network carriers are broadly speaking making sensible moves. They are endeavouring to ease the burden of legacy working practices, though the negotiations with organised labour are challenging, to say the least. Network carriers appear, mercifully, to be maintaining capacity discipline; with management having scant regard for market share measures. Attention is rather focused on boosting unit revenues, buttressing fragile fare revenues with ancillary revenues where possible.

So we are optimistic; we do think the weight of probability is that the economic recovery should be better than an L-shape and we do think management teams are making progress strategically and operationally.

However, we are far from confident that there is any obvious path towards a value creating industry. We do not see any strategic move by network carriers to meet the challenges of low cost carriers on short haul routes or Gulf carriers on long haul. We expect to see the global trend of liberalisation continue to pick off those remaining pockets of regulated super normal profitability. Meanwhile competition authorities delay the pace of joint ventures and consolidation. Nationalism stands in the way of true cross border consolidation. Finally increasing environmental awareness looks set to combine with battered government budgets to bring a swathe of increased taxation and charges on the industry, no doubt all wrapped in fine biodegradable recycled green packaging.

Enjoy the book. Nawal's books always offer an insightful look at our industry and interesting ideas from other industries too.

London, UK
December 2009

Foreword

Samer Majali
Chief Executive Officer
Gulf Air

Between the time of Nawal Taneja's last book, *Flying Ahead of the Airplane,* and this sixth book in his series, the global airline business has lost in excess of US$10.4 billion in 2008. At the time of writing, IATA is forecasting a loss of US$11 billion for this industry in 2009, losses which I anticipate will continue into 2010.

Given such persistently dismal financial performance and against the backdrop of difficulty in finding credit to continue funding this capital intensive industry, questions about the sustainability of our business as a whole are becoming a popular topic of discussion.

On the other hand, it is opportune to remind ourselves that over 5.5 million[1] workers are employed directly in this industry worldwide, with a turnover of more than US$1 trillion. If aviation was a country, it would rank 21st in the world in terms of Gross Domestic Product (GDP), amounting to generating US$425 billion, which is considerably larger than some members of the G20. Aviation, its supply chain and the spending of employees in these businesses support more than 15 million jobs and US$1.1 trillion of GDP worldwide. Taking into account the additional industries that depend on air transport, these figures are even larger.

Having said that it is not an excuse to all sit back and wave our hands in dismay. Taking a leaf from the other industries which Nawal did, there are lessons to be learnt; it is time to take a long hard look into this mature yet somewhat inefficient industry.

A major impediment to our longer term financial stability relates to Aeropolitics. An example is airlines' difficulties in achieving

1 *Aviation, The Real World Wide Web* by Oxford Economics.

consolidation by cross-border investment. Consolidation in our industry varies from mergers and acquisitions, to the different forms of alliance and partnership. Although the airline industry is a leader in the latter, through code-sharing services and other cooperative arrangements, the consolidation process by and large has been difficult due to the ownership and control rules, often requiring governments' approval. Given the current climate of increasingly more protectionism, this may further hamper the dire need to synergise airline businesses and operations.

Associated with this are the all familiar bilateral constraints which restrict commercial freedom for airlines. Nawal aptly surmised that the elimination of such regulatory constraints will enable airlines to achieve greater efficiency and optimal organizational structures. It is refreshing to note that the International Air Transport Association's (IATA) Agenda for Freedom Initiative has moved forward, with Chile, Malaysia, Panama, Singapore, Switzerland, the United Arab Emirates (UAE), United States and the European Commission (EC) signing a Multilateral Statement of Policy Principles regarding the implementation of bilateral air service agreements. These countries represent some 60 percent of global aviation. The task in hand is to quickly ratify more countries into this initiative.

Moving away from structural changes, another sacred cow to slaughter is governments' ongoing target of airlines as a convenient revenue collection mechanism. A case in point is the plan by the United Kingdom (UK) to increase its departure tax on air passengers from 2010, touted as a green initiative. This, together with the other charges will constitute about 15 percent of the average ticket price and there are other new waves of taxation in the pipeline.

As an airline executive these issues are very close to my heart. There are two particular points that I would like to share. First, on aviation's role on climate change: in reality the aviation industry is among the smallest polluters, around 2 percent of global CO2 emissions, and we are doing more than others to reduce emissions. Any discussion about carbon costs should recognize the benefits that air transportation brings to many worldwide; responsible policy should work towards a sustainable balance between these positive impacts and the cost inherent in future growth. These costs, through levies and taxation, must be ploughed back into the industry and for genuine carbon reduction initiatives and/or research and development, notwithstanding meeting the various carbon targets set. Governments should give their full backing and support to International Civil Aviation Organisation (ICAO), as the appropriate international body to set and administer specific standards and targets for aviation CO2 emissions. ICAO is the right vehicle to coordinate a proper global response to the issues of aviation pollution.

The other aspect is the treatment of airlines as cash cows which is inequitable and unsustainable, considering that we are only part of the equation in this industry but often regarded as the solution. As an industry we are all in this together!

In essence, while recognizing the important role the global airline industry plays in linking economies and societies in ways that are practically irreversible, the impediments to its sustainability must ease. These stem from the inherent inefficiency of the complex and outdated structure of government ownership and control rules, restrictive regulations in bilateral agreements as well as political intervention often hidden behind legitimate causes such as climate change control. Increasingly the price to pay for our industry's complexity and sluggishness is detrimental to survival. The call for action is now. Governments must un-tie the hands of airline managements so that they can innovate along the lines suggested by Nawal in this book using the best global business practices.

Bahrain
December 2009

Foreword

Hussein Massoud
Chairman and Chief Executive Officer
Egyptair Holding Company

Like most other industries, the airline industry has been heavily affected by the economic crisis that the global economy has been suffering since last year. Whether the current recession be L, U or W-shaped and whether recovery be realized in 2010, 2011 or even 2012, the rule is that the global economy is to revive. However, the real challenge for the airline industry in the course of this recession is the ability of the airlines to navigate safely throughout the storm until reaching recovery. Withstanding the high headwind and crosswinds tending to drift airlines from their targeted tracks towards profitability and growth requires from airlines a high degree of agility. Airlines' management must be sufficiently alert so as to anticipate or at least sense upcoming changes and rationally respond to them in a timely manner. In seeking a way out from difficulties encountered by the airlines, management needs not to be bound by conventional solutions and inherited constraints. The need for agility in this industry is not confined to intervals of crises. The high dynamicity of the airline industry and the nature of the challenges it is confronting imply the need for agility as a characteristic of the airline management in order to assure sustainability of business efficiency.

The soaring of fuel prices in 2007/2008 had its impact on the financial results of all airlines. This led airlines to enter into fuel hedging arrangements with various selections of the covered amounts, thus running—at different levels—the risk of becoming exposed to a financial burden that could have been escaped. The risk became a reality when later a downturn of fuel prices took place, demonstrating the importance of right-sizing the risk when taken. The high sensitivity of fuel prices to political and economic interactions can lead one to predict that fuel prices will remain a challenge, which this industry has to be ready to cope with from time to time.

Current overcapacity, augmented by the projected volume of deliveries of wide-body aircraft (leading to a mismatch between supply and demand), forms another challenge to the industry and is likely to affect competition in long-haul markets.

Low cost carriers have and will continue to acquire some of the market segments that previously belonged to legacy carriers. They are deviating from their initial business models with the aim of attracting more segments and increasing their market shares.

Regulatory changes also are expected to bring to the playground new concerns and considerations to this already complex industry. Liberalization of air transport between the US and the EU will certainly reflect on the global air transport industry. Likewise, the implementation of the articles of the European Community Treaty, allowing cross ownership among EU countries as well as the designation under bilateral agreements with non EU countries, will also induce effects on the industry. The upcoming enforcement of emission rules and the evolving emission trading schemes bring to reality a new dimension and constraint that never existed before, thus imposing an additional financial burden on this industry which already has the thinnest profit margin.

The trend of cross border consolidation and acquisitions associated with granted anti-trust immunities is perceived to vary the map of market shares leaving stand-alone carriers in a difficult position.

One of the contributors to the operating cost of legacy airlines is the charges imposed by GDS providers for distribution of inventory data through their networks. The burden of these charges unfairly cuts down the thin profit margin resulting from passenger transport operations, particularly in low yield sectors, and requires revisiting this mutual service provider-client relationship.

In "*Looking Beyond the Runway*", Dr. Nawal Taneja is providing a professional insight of the new changes and challenges that the airline industry is undergoing through a deep and thorough analysis supported by factual examples. In Chapter 4 of his valuable book, he addresses in a comprehensive manner the need for innovation in airline business and the venues for its realization. He further provides examples from industry and case studies from outside the airline industry.

On the side of opportunities, air transport remains to have the advantage of being the safest mode of transportation compared to surface transportation. This fact provides an opportunity for short-haul air transport to replace, in certain markets, other modes of transportation. Emerging markets in various regions of the world are valuable opportunities. Partnership, in form of code share and alliances, has widened the domain of beneficiaries from these opportunities in lieu of being limited to traditional traffic right owners.

The spreading use of Internet worldwide creates the opportunity for airlines to increase the share of direct distribution through Internet booking engines at the expense of GDS distribution, and hence an opportunity to reduce significantly distribution costs.

The reader of Dr. Taneja's book *"Looking Beyond the Runway"* will truly enjoy the deep insight offered. It is a call for farsightedness that should not be missed by airline executives.

Cairo, Egypt
December 2009

Foreword

Robert McGeorge
General Counsel
International Air Transport Association

Looking Beyond the Runway is the sixth of Dr. Taneja's books for airline practitioners. As with his previous books, he analyzes the turbulent, challenging and rapidly changing economic environment in which airlines operate, and provides insights for the airlines that must compete in this environment.

As IATA's General Counsel, I focus primarily on the legal and regulatory environment in which airlines operate; but there is an obvious overlap between issues of interest to airline executives and airline lawyers. I will seize this opportunity to comment on a few overlapping issues of particular interest.

In this book, Dr. Taneja addresses airline ownership and control, their access to foreign markets and their ability to enter into collaborative ventures. In an ideal world, airlines might enjoy the freedom to operate like most other business enterprises – acquiring equity capital from global investment markets, establishing global brands and entering markets with the greatest economic potential. Under current legal regimes, however, airlines face restrictions in all of these areas, severely limiting some aspects of innovation discussed in this book.

As an advocate for the international airline industry, IATA has supported liberalization of national laws on foreign ownership of airlines. Turning to one recent and positive example, seven countries and the European Commission (collectively accounting for a majority of international airline traffic) responded to our *Agenda for Freedom* initiative by signing a Statement of Policy Principles, which contains, among other measures, a political commitment to waive nationality clauses in their bilateral air services agreements that restrict ownership of their bilateral partners' airlines.

Airline lawyers who have encountered prohibitions on foreign ownership and control have devised creative legal strategies for obtaining many of the efficiencies that other industries obtain through mergers and acquisitions. By creating the legal structures for lawful collaborative arrangements, such as codeshare agreements, alliances and minority investments, airline lawyers have helped to expand airline networks, to create additional virtual online services for millions of passengers, and to enable airlines to allocate resources more efficiently.

Dr. Taneja has stressed the importance of controlling costs. Most of the credit for airline's remarkable success in reducing controllable costs must go to business executives. IATA attorneys, and our colleagues in airline member legal departments, have worked diligently, but with less success, on the legal environment that is responsible for a significant portion of airline costs.

One obvious area of rapidly escalating costs is taxes and fees that governments impose on airlines and their passengers. For example, these fees and charges amount to about half of the total cost of a $300 roundtrip Montreal/Washington ticket. The airline industry remains the favorite source of funding for new governmental programs, ranging from the protection of the environment, to supporting economic development in developing countries, to providing compensation for non-passenger injuries caused by terrorist interference with aircraft operations. Unfortunately, we cannot claim much credit for reining in these costs.

Private businesses also treat airlines like "cash cows." Like most businesses trying to survive turbulent times, airlines face fierce competition when they sell their services. Unlike most other industries, however, when airlines buy goods and services they often have no option other than to negotiate with monopoly or oligopoly sellers – jet fuel suppliers, airports, ground handling companies, air navigation systems, and global distribution systems to name a few.

In some cases, governments have mandated or tolerated monopoly suppliers when competition is entirely feasible (e.g., ground handling services at most airports). In other cases, the monopoly supplier enjoys a natural monopoly (e.g., airport or airline navigation services). With mixed success, IATA has advocated that privatization should not result in unchecked private monopolies. If competitive forces cannot effectively discipline the prices of those who sell their goods and services to the airline industry, shouldn't governments discipline those prices through effective and efficient regulatory systems designed to replicate market forces and incentives to the extent possible?

Finally, I must agree with Dr. Taneja's recognition of the importance of good market intelligence data in responding efficiently and effectively to market conditions. Among our business intelligence products, IATA offers airline ticketing data.

Almost every other entity that "touches" airline reservation and sales data also recognizes the value of this asset – including passengers, travel agents and global distribution systems. Some of them claim exclusive rights to control the dissemination of that data.

This is not the place to argue the merits of conflicting claims of rights to control the dissemination of this data – although they raise interesting legal issues at the intersection of intellectual property and antitrust/competition law. It seems appropriate, however, to recognize that if other entities in the airline distribution system gain the sole right to disseminate airline reservation and ticketing data, the airlines will not fully utilize the power of good market intelligence.

If there is one basic theme to these examples of overlapping business/legal issues, it is that airline executives and attorneys must work closely, diligently and creatively to survive (and hopefully to thrive) in this environment. We should advocate changes in legal and regulatory systems that will give airlines the rights, responsibilities and flexibility of other international business that are trying to survive these turbulent times. We should not be too pessimistic, however, or put all of our hopes on legal reform. Dr. Taneja has analyzed the success of airlines that have overcome these challenges, and ventured outside the industry to provide insights into creative ways of doing business that are available to us under the current legal and regulatory environment. Airline attorneys and business executives must work together to create success stories that compare favorably to the examples Dr. Taneja describes in this book.

Montreal, Canada
December 2009

Foreword

Gary R. Scott
President
Bombardier Commercial Aircraft

Dr. Nawal Taneja and I share a fundamental view that the time is now for the aviation industry to step-up to our challenges and take control of the future. While good management may deliver short-term fixes, visionaries backed by good strategies hold the key to a more sustainable, long-term prosperity. We need to create and shape our future!

Dr. Taneja is such a visionary, and the next seven chapters of *Looking Beyond the Runway* will convince you that change is necessary, re-invention is vital. This book is for me and it's for you – leaders in the aviation industry. He brings home the point that *we are the change*.

At Bombardier, my team and I have been tasked to be the "agent of change" and introduce the all-new technology "game changing" CSeries aircraft program.

In fact, in commercial aviation you need look no further than CSeries for a tangible example of the "new normal" that Dr. Taneja articulates in *Looking Beyond the Runway*. When we set out to launch CSeries we examined what is, and envisioned what should be. We quickly realized that a new product needed to take advantage of all new available technology to offer maximum flexibility, a dramatic improvement in cost over the aircraft life cycle and a breakthrough in environmental impact. Above all, we focused on our customer and our customer's customer and we are determined to address every need, every expectation.

A recent survey by Ascend confirms that the top two concerns of 192 professionals from 90 airlines are fuel price instability and declining revenue and yield. Yet at the same time, passengers (our customer's customers) continue to be more demanding not only in terms of the on-board experience but also in terms of choice.

The challenge for an aircraft manufacturer is to deliver on both fronts.

In the 1990s, the Bombardier CRJ changed the game for airlines and passengers by introducing reduced travel times and more choice through the development of high frequency service connecting cities to hub and spoke networks. Not long ago, skeptics didn't believe anyone would want to fly across the ocean in a single aisle aircraft. Today, several airlines offer non-stop trans-Atlantic flights to smaller cities and can extract a premium for the convenience.

The CSeries is another game changer – greener, flexible, more comfortable and purpose-built to make money for airlines in the key 100- to 149-seat market space. The CSeries promises to enable yet another dimension for airline growth by providing visionary airline management a tool for real network optimization instead of traditional adjustments with the last generation of aircraft designs.

Dr. Taneja points out that resisting change because of cost is perilous. The CSeries investment will run well over $3 billion for Bombardier and our partners. Some might characterize this program as a "gamble"; I would say that is true of all new game changing programs. We accept this challenge with confidence and along with our other stakeholders understand that it is not without risk, but a risk worth taking.

The potential payback more than justifies the investment plus it will propel us into the future, continuing our leadership and brand growth. Sounds like something out of Dr. Taneja's playbook!

Montreal, Canada
January 2010

Foreword

Robert Solomon
Senior Vice President and Chief Marketing Officer
Outrigger Enterprises Group

Nawal Taneja does the industry a great service by searching both within the global airline community and beyond for insights and best practices. Over the years, he has demonstrated clearly that the differences among airlines are as great, or even greater than, the apparent similarities. While there are some striking examples of success, there are only a few cases of management-led transformation. The imperative for strong, focused leadership, and the need for a positive and growth oriented organizational culture, resonate throughout.

Nawal navigates us deftly though "good" strategies that failed (for example, the Saturn car company), and improbable strategies that have worked (Allegiant, LAN and Air Asia come to mind), reminding us of the primacy of execution and the limitations of conventional thinking. Cases like Zipcar, Zappos and, previously, Zara, help to stretch the horizon for innovation.

For some of the most vital lessons in this fast-paced work, we must read between the lines. Whether or not the specific cases are relevant to one airline's leadership or another's, how many airline leaders can say they are truly focused on the consumer, present and future, or that their organization is pushing the envelope to take advantage of the best and most effective means of going to market, in markets that are transforming at unprecedented speed?

How many have embraced the concept that *all* travel has become discretionary, that growth in revenue and profit can indeed come from the now-dominant and still faster growing leisure segments, and that *relevance* is the key to earning consumer preference?

Successful innovators must be more inventive than inventors. They scan the horizon for models, technology, insight and leadership that has transformed other businesses and categories, and apply it smartly, out of context. The best "out of the box" thinking often comes from another box. But how can airline industry leadership bring these lessons into their own organizations, when they and their beleaguered colleagues spend their business days and nights grappling with the endless challenges of their own industry environment?

In the airline sector of the travel industry, content consists of pricing and schedules. In the hospitality sector, where Outrigger Enterprises Group has a growing portfolio in Hawaii and the Asia-Pacific region, we strive to provide a genuine and differentiated customer experience. While the cultural context varies, employee commitment is always rooted in a shared set of fundamental values that travel very well. When our guests can feel the difference, they become our most effective marketing channel, and we apply leading edge technology to convey the relevant story to the right customer at the right time. Although our booking usually comes after the airline transaction, we work very hard to inspire the decision to fly.

Consultants are always ready to help with "transformation" and "change management" programs, but the leadership team has to make the decisions and the culture has to support the buy-in. This is where Nawal Taneja brings significant value to the airline industry—by challenging, provoking, and assisting the process of looking across, beyond, and ahead of the flight path.

Honolulu, USA
November 2009

Foreword

Andrew B. Steinberg
Partner, Jones Day
(Former Assistant Secretary of Transportation for Aviation and International Affairs, U.S. Department of Transportation; and Chief Counsel, U.S. Federal Aviation Administration)

When Congress deregulated the U.S. airline industry in the late 1970s, it was wiser than we now seem to give it credit for. The Federal Aviation Act as amended required the U.S. Secretary of Transportation to ensure that "efficient and well-managed air carriers to earn adequate profits and attract capital," that "consumers in all regions of the United States, including those in small communities and rural and remote areas have access to affordable, regularly scheduled air service," and that U.S. airlines enjoy a competitive position of least "equality" with foreign air carriers. Title 49, United States Code, Section 40101. Yes, in case you are wondering, this law is still on the books. But we'd have to give ourselves failing grades on each of these measures.

I have long felt that the simplest explanation for this failure lies in our reluctance to pursue policies consistent with these explicit objectives. If three decades of deregulation has taught us anything, it is that we cannot satisfy consumers when the only thing we seem to really care about (other than safety) is treating air transportation as a commodity and keeping airfares at 1978 levels. So to have a profitable industry that better serves the interests of its customers over the long term, it seems to me that we don't need new laws; we just need to go back to Congress' original goals in applying the laws we already have. As a matter of aviation policy, this means moving away from a singular focus on lowering prices to consumers, the guiding principle to date.

This fundamental misapplication of our objectives created not a mere academic problem but a big commercial one that has largely eluded a solution. Over the years mistaken policy has produced a visceral and foolish opposition to almost any form of consolidation by network airlines, harebrained attempts to legislate customer service (such as current bills in Congress that would micromanage air carriers in their handling of ground delays), assaults on the networks themselves (read: "fortress hubs"), and most bizarre of all, proposed rules to prohibit large airlines from matching prices set by smaller airlines. By pursuing policies that were designed to ensure major U.S. airlines had no pricing power, and that fares would always be low, we engineered a perverse outcome that has harmed consumers in a more lasting way: As the legacy carriers were forced to slash domestic capacity beginning in 2008 (due to sky high fuel costs that they could not pass on to their passengers), prices began to go up, not down. And despite fewer flights, we saw abysmal customer service, increasing delays, and a frightening reduction in service to rural communities. (In a hub and spoke system, airlines pull out of their thinnest markets when times get tough.)

We all know that the rules of capitalism were never suspended for the passenger airline business. Like other companies, airlines that perennially lose money cannot afford to reinvest in their core businesses, much less innovate or improve them. Instead, they focus on cutting costs and just staying afloat. Even when they make profits, if the long term prospects for their shareholders remain bleak, they are wise not to reinvest the money in new capital assets. Sadly, eight years after the shocks of the World Trade Center attacks, none of our largest network carriers has replenished its fleet. This situation is disastrous for U.S. airlines competing in an increasingly global marketplace against the likes of Emirates, Singapore Airlines and Lufthansa. In fact, our legacy carriers have fallen far behind their foreign rivals on every measure of success: profitability, average fleet age, market value, and customer service. Instead, we are forced to excel in the creative use of the bankruptcy laws.

But what is the solution? As one alternative, some have proposed re-regulation of fares and markets, but not too many observers believe this step is desirable (consumers would clearly be worse off) or practicable (who stops flying where?) in a world where most other countries are rapidly removing the last vestiges of regulation. As a another alternative, Wall Street investors say that the real solution is to permit the airlines to consolidate –

through mergers, asset sales, or exit by failed firms -- just as other deregulated industries have done. (The airline industry today is still as highly fragmented as it was after being deregulated three decades ago.) Proponents of this idea point out that to have sufficient competition over prices, we don't need five network air carriers flying largely duplicative route structures today any more than we would need five wireless phone companies, five Internet search engines, or five express package carriers.

Prof. Nawal Taneja's latest book, *Looking Beyond the Runway*, provides a third alternative, and will provoke healthy thought and necessary debate over what really ails our airlines and how to cure them. By treating the shortcomings of the legacy airline industry as a simple failure of *innovation* -- and then offering, in systematic detail, case studies from other industries -- he makes a persuasive case for viewing the airline crisis in more classical business terms. He explains how innovative ideas in other sectors of the economy as diverse as hamburger restaurants (you have to read the book to know why I don't refer to these businesses as "fast food"), internet search engines, and clothing manufacturers could be applied to airlines. And he gives some insight into those airlines that appear successful at innovation. One is left with the feeling that improvement is really possible.

The question I personally get most often when new acquaintances learn I have worked in and around the aviation and travel sectors for a long time is, "Why can't airlines behave like other businesses? " Prof. Taneja's book answers the question by proving that they can and, probably, should. *Looking Beyond the Runway* is a welcome contribution to the business literature on airlines and an important book for practitioners in the aviation industry to read.

Washington, D.C., USA
November 2009

Foreword

Junku Yuh
President
Korea Aerospace University

The current global economic downturn has made an enormous impact on the global aviation industry: Passenger and cargo traffic, and asset values have declined; and the fundamental structure of the aviation industry has changed. These changes are dramatic in their depth and breadth, and stem primarily from demographic, social, technological, political, regulatory, and economic trends. However, while these changes have produced challenges for the aviation industry, they could also produce opportunities if the suppliers in the air travel value chain would change their business models and government policy makers would create an environment that makes it easier for aviation service providers to introduce innovation. The need for air transportation in this increasingly global world will continue to grow, perhaps, even much more in developing economies. In fact, analyzing the working profile of the global aviation industry for viable improvements is one of major focuses of the College of Aviation and Management in our university, Korea Aerospace University (KAU).

Dr. Taneja makes an important point in this book, *Looking Beyond the Runway*. It is true that the global airline industry is constrained by many factors, such as varying degrees of government control in different regions and the labor, capital, and fuel intensive nature of the industry itself. However, the operators in the aviation industry must learn to innovate in the presence of these constraints. This point is well aligned with what we emphasize in the classroom at KAU, where students learn how to optimize the system within the realities of the marketplace. Through the aviation education programs, we assure that the students will contribute to efforts in making aviation a leading industry when they work in the real world after graduation.

Another point that Dr. Taneja has made in this book is that it has a great value for the aviation industry to learn from the best global practices in operations, logistics, and other areas. We at KAU have already introduced this concept to aviation education programs. We start with cross-functional integration within the aerospace field. For example, in the study of the current and future aviation infrastructure, such as airports and air traffic control systems, we include a study on how a new aircraft is designed. We move even further by looking at emerging considerations such as the protection of the environment. We also teach and research about the need of integrating on a system basis and considering the interest of all stakeholders in the aviation industry, such as users, service providers, policy makers, regulators, and labors.

As Dr. Taneja mentions in both Chapters 4 and 7, advanced airplane technology has always made a major impact on the global airline industry. This is true and no one would disagree that a highly trained and skilled workforce is one of the key contributors to the advancement in aviation. As the aviation industry will face more challenging and complicated issues due to higher competition and additional constraints such as regulations regarding the protection of the environment, the industry will require people with more advanced and analytics-oriented knowledge. KAU constantly update aviation education programs to equip the students with the skills needed to pursue successful careers in various sectors of the aviation industry. Dr. Taneja's book provides additional insights into the areas where we can make our education programs even more comprehensive and the value of the students trained by our programs will be recognized by the global aviation industry. I am sure that many other aviation education programs in the world will also receive the same benefit or even more from this book.

Goyang City, Korea
November 2009

Acknowledgements

I would like to express my appreciation for all those who contributed in various ways, especially, Angela Taneja, an experienced analyst of best global business practices, Dr. Dietmar Kirchner (formerly with Lufthansa and now a Senior Aviation Consultant), and Rob Solomon (Senior Vice President and Chief Marketing Officer at the Outrigger Enterprises) for discussions on challenges and opportunities facing the global airline industry and related businesses.

The second group of individuals that I would like to recognize include, at: Air Canada—George Reeleder; Airline Monitor—Edmund Greenslet; Airline Intelligence Systems—Stephen Johnston; Air Transport News—Kostas Iatrou; @aquila—David Palmieri; Boeing—Fariba Alamdari; Bombardier Aerospace—Jerome Cheung, Chuck Evans, and Philippe Poutissou; Centre of Asia Pacific Aviation—Peter Harbison and Binit Somaia; Continental Airlines—Chris Amenechi, Greg Hart, Scott O'Leary, Alex Savic, and John Slater; Copa Airlines—Pedro Heilbron, Joe Mohan, and Marco Ocando; Delta Airlines—Chul Lee; Expedia—Greg Schulze; fly.com—Brian Clark; Forrester Research—Henry Harteveldt; GOL—Fábio Chagas Sanches; IATA—Brian Pearce and Patricio Sepúlveda; ICAO—Eduardo Chacin; LAN—Gisela Escobar and Enrique Cueto; Lufthansa Airlines—Jens Bischof, Nico Buchholz and Christoph Klingenberg; Royal Bank of Scotland—Andrew Lobbenberg; Southwest Airlines—Adam Decaire, John Jamotta, Lee Lipton, and Pete McGlade; SITA—Jim Peters; TAP Air Portugal—Fernando Pinto; Turkish Airlines—Ahmet Bolat and Temel Kotil; Ypartnership—Peter Yesawich; Syncrata—George Gendron and Michael Kron; WestJet—Hugh Dunleavy. In addition, there are Barry Humphreys (formerly with Virgin Atlantic) and Scott Nason (formerly with American Airlines).

Third, there are a number of authors whose work and ideas have been referenced numerous times in this book. They include James Allen, Ron Alsop, Chris Anderson, August Joas, Leonard Berry, Ram Charan, Geoffrey Colvin, John R. DiJulius III, David Edery, Judy Estrin, Paul Flatters, Dale Furtwengler, Rowan Gibson, John Gourville, Linda Gravett, Jill Griffin, Kara Gruver, Ronald Haddock, Shep Hyken, Ron Harbour, Jeff Jarvis, Katherine Jocz, John Jullens, Tom Kelley, Parag Khanna, W. Chan Kim, Vijay Mahajan, Renée Mauborgne, Micheline Maynard, Rafi Mohammed, Ethan Mollick, Dev Patnaik, John Quelch, Darrell Rigby, Larry Selden, Peter Sheahan, Dilip Soman, Peter Skarzynski, Christopher Steiner, Alex Taylor III, Bruce Temkin, Robin Throckmorton, Vijay Vishwanath, and Michael Willmott.

Fourth, there are a number of other people who provided significant help: at the Ohio State University— Gary Doernhoefer, Josh Friedman, Alex Holmes, Robyn Litvay, and Jim Oppermann; and at the Ashgate Publishing Company (Guy Loft—Commissioning Editor, Kevin Selmes—Production Editor, Mike Brooks—Manuscript Reader, and Luigi Fort—Senior Marketing Executive).

Finally, I would also like to thank my family for its support and patience.

Chapter 1
Outlining the Chaos, Evolving Strategies, and the New Normal

The Chaos

Dealing with chaos is hardly a new phenomenon for the global airline industry. During the past four decades alone, it has survived numerous oil price shocks, significant changes in government regulatory policies, new technologies (ranging from aircraft with advanced capabilities to the introduction of the Internet and related businesses), competition from new categories of airlines and aviation, varying lengths and depths of economic cycles, the events of September 11, 2001, diseases such as SARS and H1N1, two wars in the Gulf region, infrastructure constraints and development, and the impact of changing environmental regulations. While the global airline industry has managed to survive, albeit with a growing tendency towards bankruptcies and mergers, it has not managed to earn a sufficient return on the investment deployed to cover its average weighted cost of capital on an ongoing basis. Figure 1.1 shows the net profit margin for the global airline industry for the past 30 years. There are two out-of-the-ordinary points to notice in this chart. First, even during the best years, the highest net profit margins achieved were only about 3 percent. Second, on a cumulative basis, the global airline industry did not break even. For the 30 year period, the industry posted a slight negative cumulative net profit margin, a phenomenon unseen in any other industry.

Although the airline industry has always faced variable headwinds, starting in 2007, the global airline industry began to face, not only one, but a confluence of strong forces, creating an almost perfect storm. While each force, itself (an enormous and sudden increase in the price of fuel; a deep recession; an unusual financial crisis, affecting both the credit market and the volume of premium travel; changes in foreign currencies; acceleration in the use of social technology to communicate the travel experience of passengers), represents a major challenge for this industry (with already paper-thin profit margins), their confluence is finally producing a herculean challenge, from a short- and long-term perspective.

From a short-term perspective, as of 15 September 2009, the global industry is forecast to lose US$11 billion during 2009, on top of the US$10.4 billion losses incurred during 2008. (The US$10.4 billion loss in 2008 has now been revised to US$16.8 billion to reflect "restatements and clarifications of the accounting treatment of very large revaluations of goodwill and fuel hedges.")[1] The major contributor to the losses in 2008 was the sudden and enormous increase in the price of fuel, climbing from about US$60 a barrel for crude oil at the beginning of 2007

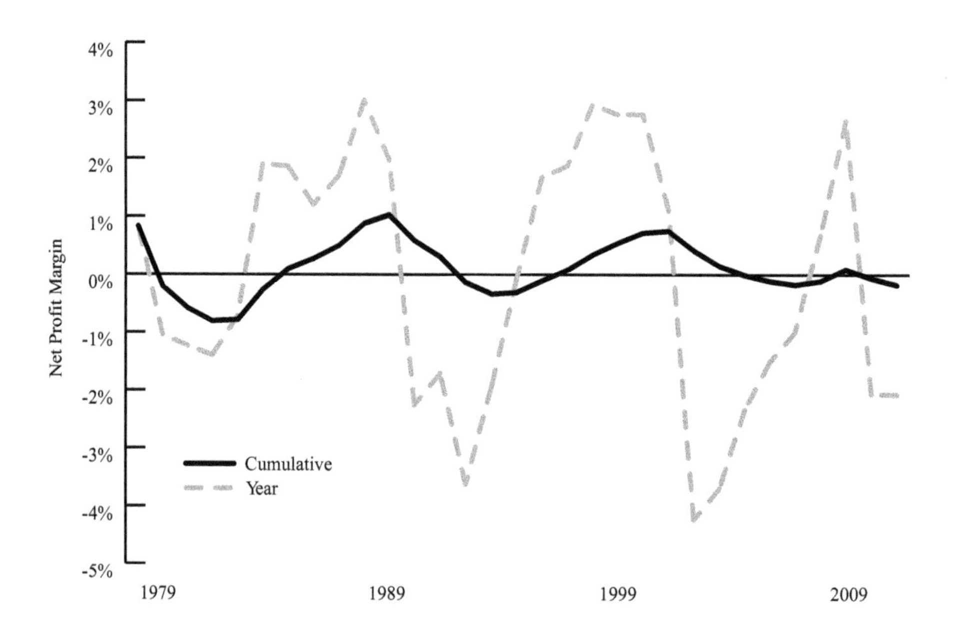

Figure 1.1 Net Profit Margin: The Global Airline Industry
Source: The Airline Monitor, June 2009, p. 24.

to US$147 a barrel in mid-July 2008. The major contributor to the losses in 2009 appears to be mostly on the revenue side, resulting from both reductions in traffic and in yields. The significant reductions in both the volume and the prices of premium travel are particular concerns. See Figure 1.2. During the first quarter of 2008, premium passengers (those traveling in Business Class and First Class) accounted for nine percent of the total passengers in international markets, but they generated 30 percent of the passenger revenue.[2] In the first quarter of 2009, revenue from premium travel declined 33 percent. In the second quarter of 2009, it declined 41 percent relative to the same period in 2008. While most of the expected losses in 2009 are due to a reduction in traffic and yield, some losses are due to an increase in the price of fuel.

If we assume the first element of this chaos to be the reduction in passenger (and cargo) traffic and the reduction in premium traffic and premium fares, then the second component would be the increasing level of incursions being made by low-cost carriers worldwide. Take AirAsia as just one example. During the second

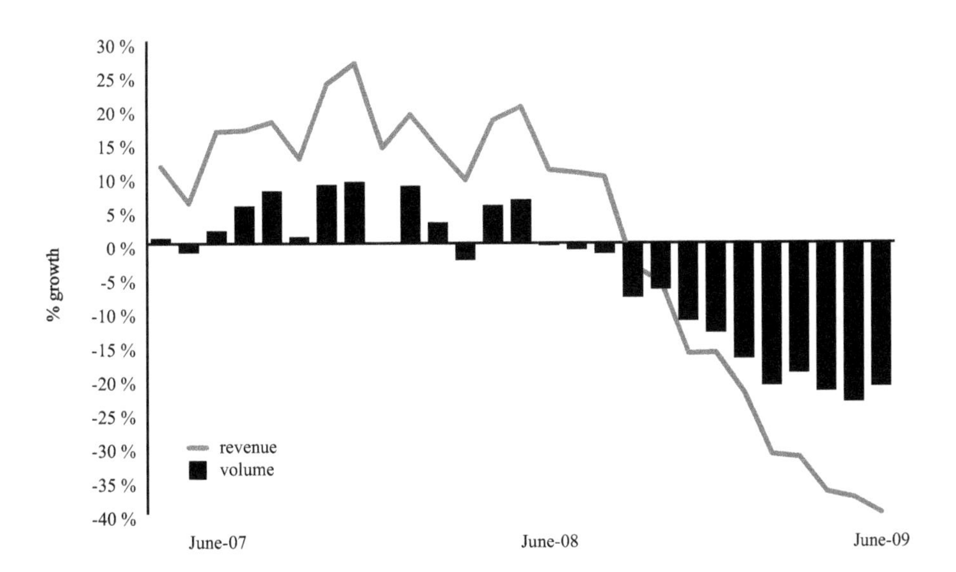

Figure 1.2 Total Premium Traffic and Revenue Growth (Year-on-Year, May 2007 to June 2009)
Source: International Air Transport Association.

quarter of 2009, AirAsia posted an increase in capacity and traffic of 22 percent and 24 percent, respectively, as well as a 15-fold increase in net profit over the second quarter 2008.[3] Similarly, low-cost carriers have been performing well in other parts of the world, Ryanair and easyJet in Europe, Air Arabia in the Middle East, WestJet in Canada, and GOL in Brazil.

There is, of course, no single low-cost airline business model. At one end of the spectrum are airlines operating with a single type of aircraft, with high-density, one-class configuration, serving point-to-point markets, with traffic that is more stimulated (for example, diverted from other modes of transportation) and less diverted from existing airlines. Ryanair could be an example using such a business model. At the other end of the spectrum might be an airline with a hybrid business model using multiple types of aircraft in a broad spectrum of markets. Air Berlin might be an example of airline using such a model. In between the two ends of the spectrum are airlines with very different models, such as those with code-share agreements with network carriers. Examples include Virgin Blue with Delta, jetBlue with Lufthansa, and GOL with American. In general, numerous different business models of low-cost carriers are proving to be more durable compared to the legacy carriers, enabling low-cost carriers to become much more aggressive in expanding their operations and in their pricing policies. Just consider the experience of Azul in Brazil. Having been launched in December of 2008, Azul became the third largest carrier with a domestic market share of just under 5 percent in July 2009 (with TAM and GOL, each having just over 40 percent).[4]

For a variety of reasons, including the major shift resulting from the "trading down" phenomenon, at least in the developed economies, the market share of the low-cost carriers could increase significantly. EasyJet, for example, could easily siphon off a significant portion of not only British Airways' traffic in London, but also Air France-KLM's traffic in Paris. Moreover, since the low-cost carriers offer a much greater percentage of their services in the point-to-point markets, large hubs such as Frankfurt, Heathrow, Hong Kong, JFK, Los Angeles, O'Hare, Paris (CDG), São Paulo, Singapore and, Sydney could see a significant impact on their operations. Some airports may, in fact, be forced to change their business models with respect to the diversification of their revenue base and the satisfaction of their customers. Finally, the operations of the low-cost carriers are no longer restricted to their own regions. The intercontinental markets, once thought to be impenetrable by low-cost carriers, have begun to show economic viability, exemplified by the experience of AirAsia X, Jetstar, and V Australia.

Evolving Strategies

As stated above, airlines are not unaccustomed to changing capacity and pricing policies to adapt to the changes in demand. Figure 1.3 shows the results of the global airline industry's actions to harmonize capacity and demand during four cycles in the past four decades. In general, airlines have done a reasonable job, given the inherent constraints relating to the airline industry.[5] Examples of constraints include long-term leases on fleet, the "use-it-or-lose-it" rules relating to slots at key airports, and, in some cases, the governments' policies to maintain service for social reasons.

Given the severity of this downturn that began at the end of 2007, a number of legacy airlines are fighting for their survival and, as such, have begun to implement changes more rapidly and more intensively. The first major challenge began to appear in 2007 when the major US network airlines began to face increasing levels of competition from low-cost, low-fare airlines. Their initial response was to shift capacity from domestic markets to international markets where competition was limited. Next, the price of fuel began to increase at an exponential rate. During 2007, the US airlines paid for fuel, for example, an average price based on about US$70 per barrel. Within six months, the price of oil increased to over US$140, a level double the average price during 2007. Such an exponential increase in the price of fuel caught the global airlines industry off guard, leading to varying degrees of reductions in capacity. Although most airlines around the world followed a similar strategy, the airlines in the US began to make the capacity cuts more rapidly and more deeply using a combination of techniques (such as the use of lower capacity aircraft and a reduction in frequency).

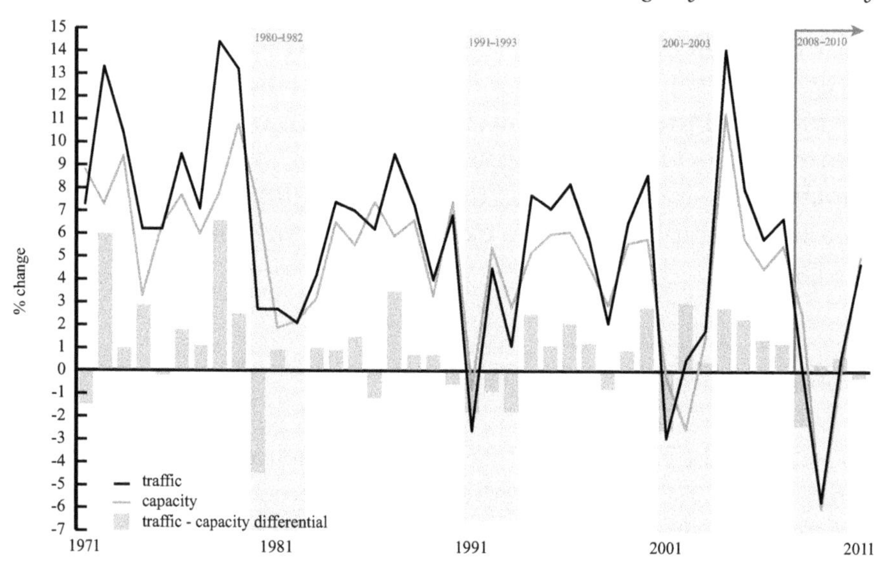

Figure 1.3 Global Airline Passenger Traffic and Capacity
Source: Airlines International, June-AGM 2009, p. 42: ICAO and IATA.

On the revenue side, airlines have been focusing much more intensively on ancillary revenue by unbundling the product and implementing à la carte pricing policies (for more information on pricing, see Chapter 5). While these strategies have increased revenue streams, they have also led to a certain degree of resentment on the part of passengers. Only a part of the resentment problem relates to the concept itself. In fact, some passengers even appreciated the à la carte pricing policies, enabling them to purchase only the product features they desired. However, the irritation experienced by some passengers relates partly to the areas in which airlines instituted charges (such as a charge for seat selection and reservation change fees) and partly to the level of the charge (such as the cost for transporting the first bag and each additional bag and, in some cases, the level of reservation change fees). Passengers can tolerate additional charges for services that lead to additional costs for airlines, for example, food served in the cabins. However, many passengers object to paying for services that do not cost airlines any money, such as which seat the passenger occupies in the economy cabin. From the airline side, on the other hand, the concept makes sense in that each seat has a different value, enabling airlines to maximize their revenue, a concept no different than the one in the sports and theater businesses.

Even though airlines have been doing reasonably well with the network, fleet, and schedule changes in general, two areas remain a major concern, (a) the reduction in Premium Class travel and the reduction in cargo traffic, and (b) the growth of low-cost carriers. The reduction in cargo traffic is the result of consumers buying less in North America and Europe that, in turn, means a reduction in the transportation of manufactured goods from emerging markets, such as China. Similarly, there is a reduction in the transportation of some perishables such as fresh flowers. Although there is no other viable alternative for transportation for perishable products, the demand is down, as people are buying less fresh flowers. However, while the reduction in cargo traffic can be considered as a cyclical phenomenon in line with the decline in world trade, the reduction in premium travel may be a lasting experience.

Despite the progress made in reducing costs and enhancing revenues using mostly conventional strategies, at least, four questions remain on the minds of traditional network airlines (leaving aside the uncertainty relating to the price of fuel and the availability as well as the cost of credit).

- If and when will the premium traffic return?
- Will consolidation move much further?
- How much more will low-cost airlines grow within their own regions and will they be successful in penetrating the intercontinental markets?
- How much of an impact will new global network carriers (such as those based in the Gulf region of the Middle East) have?

As for the first question, the continual decline since mid-2008 in premium demand in international markets served by major full-service airlines is a profound change in that the major premium market is not likely to recover fully. The reasons appear to be (a) that most businesses are controlling their travel budgets much more tightly and, (b) that many individual travelers who pay for their own tickets do not consider the Premium Class, price-service options to be valuable. The provisions of accounts that airlines have with corporations are likely to change if airlines continue to offer discounted fares that are lower than the corporate discounted fares, and airlines expand their initiatives to lower, or eliminate, some restrictive fences previously deployed to stop business travelers from being able to take advantage of lower fares available to leisure travelers.

As for consolidation, airlines have stepped up their interest in the different consolidation options. The main idea behind consolidation is to capitalize on synergies to reduce costs and increase revenue. Consolidation strategies have varied from mergers and acquisitions to different types of alliances and different levels of participation within alliances. The consolidation process has been difficult due to the existence of ownership and control rules and even alliance participation requiring government approvals. However, the code-sharing level of participation in alliances has been relatively easy and the process has enabled alliance members to offer passengers integrated services around the world while achieving some cost reductions. Higher levels of consolidation have met varying degrees of resistance from governments.

Take the case of British Airways (BA) and Iberia (IB), carriers that have been in discussions to consolidate in a way that allows them to maintain their two brands, but coordinate schedules and fares across their combined networks. While BA has had a link with Iberia for almost a decade through a minor equity stake, the link has been limited to a joint venture on some routes between the UK and Spain and the standard agreements between alliance partners. The pressure was brought about by the increase in the competitiveness of Air France-KLM, not only from their combined operations (since 2005), but also from the additional capacity available at Paris' Charles de Gaulle to expand operations relative to the capacity at London's Heathrow. Similarly, not only did Lufthansa start to grow organically, but it also acquired

Swiss International, and quickly turned it into a highly profitable operation.

Next, came the open skies regime across the Atlantic. Then the impact of the increase in the price of fuel and the decline in traffic from the recession made the need for the consolidation initiative more imperative.

Finally, the two four-way joint ventures (Delta, Northwest, Air France, KLM) and (United, Air Canada, Lufthansa, Continental) put an even greater pressure on British Airways not only to have a joint venture with Iberia (finally agreed in November 2009), but also for the two to have a joint venture with American. Such joint ventures will bring more synergies to reduce costs (through joint purchases and joint marketing programs such as sharing of airport lounges), but more important, use of the synergies to optimize the margin. This effect, however, has raised eyebrows from some anti-trust authorities suggesting that the initiatives may not be in the public interest.

As for the expansion of low-cost carriers, both older and newer low-cost carriers are demonstrating enormous staying power. Take the case of AirAsia (which will be discussed in Chapter 4). It has developed not only huge bases in Asia, but it has also begun services in intercontinental markets. Consider also the performance of Sharjah-based Air Arabia. This carrier has not only been posting positive financial results during the downturn, but the carrier now has two subsidiaries, one based in Morocco, and one based in Egypt. The group now serves 57 destinations across Europe, the Middle East, and Asia, ranging from Casablanca, in Morocco to Chittagong, in Bangladesh. The experience of Jetstar also substantiates the force of low-cost carriers. Launched in December 2005, it is now the second largest international airline in terms of passenger market share of Australia's international aviation market, passing huge carriers such as Singapore. Jetstar is now the largest operator between Australia and Japan.

As to the question of new global network carriers, while most traditional legacy carriers have been contracting out their operations and implementing strategies just to survive, there are a few airlines with different mindsets, coupled with some inherent strength, that are looking at the current chaos as an opportunity to not only expand now, but also position themselves to take further advantage of the changing marketplace after the start of the recovery. The change in the landscape has provided an opportunity for some airlines with resources to expand their operations. The Gulf region of the Middle East is one example. Using the fleet that they had on order, the strength of the network they had developed, and the liberal regulatory policies in certain markets, the Gulf-based carriers began to expand their operations. Take the case of Emirates. While the airline is suffering financially to some extent, it is less so than many other carriers because of the strength of its network. The network is highly diversified. Consequently, even with the reduction in passenger traffic to and from Dubai (as a result, for example, of the reduction in building activities in Dubai), Emirates is able to expand its operations because of the enormous connectivity network at its hub in Dubai. A case in point: Passengers in six cities in the UK can connect to nine cities in India and four cities in Australia via Dubai. See Figure 1.4.

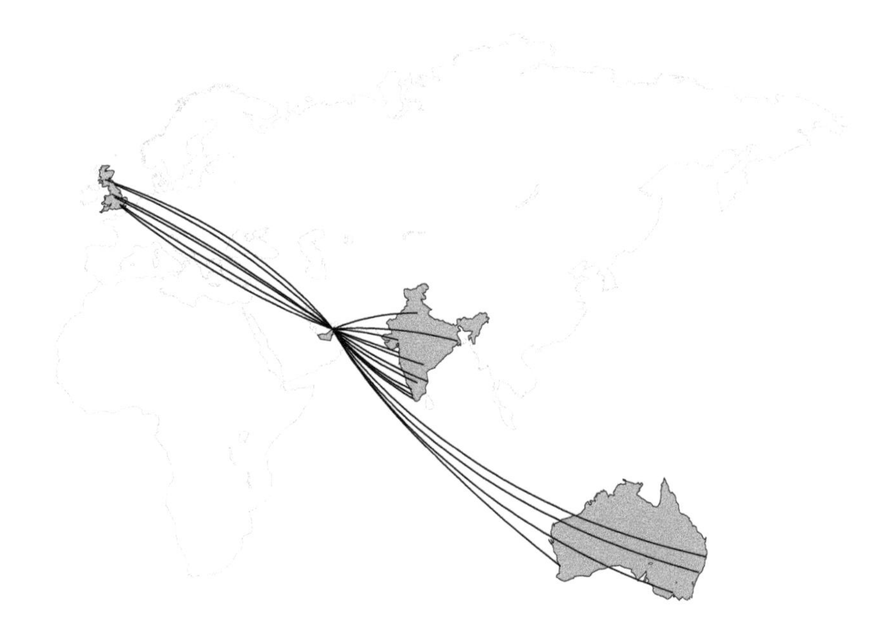

Figure 1.4 Emirates' Routes from the UK to India and Australia via Dubai (July 2009)

Source: Based on the information available on the website of the airline.

The Gulf region is doing well because of its geographic location, the unique mix of markets, not just the standard business, VFR, and leisure travel (growing investments in tourism related projects), but also pilgrimage related travel, and labor workers. The Gulf carriers are investing in aircraft, infrastructure (not just new runways, but also new terminals, new airports and even relating to ATC), and in in-flight product. They also are not saddled with legacy issues facing other carriers, for example, in old fleet, old IT systems, and old labor contracts. They will also be able to take advantage of the Premium Class travel when, and if, it returns. For these airlines, tough times mean having a slower growth rate, for example, from 15 percent down to 8 percent. Similarly, some Asian and Latin American carriers have also been doing reasonably well relative to those in the US and Europe.

The New Normal

From a longer-term perspective, and, at least in North America, it is now a widespread view within the business community that companies will no longer operate on a "business as usual" basis in the post-crisis world. The post-crisis world, at least in many western economies, will be fundamentally different for many reasons. Let us first begin with some factors that are related indirectly to airlines.

- Composition, performance, and relative importance of economies, particularly of the US, will most likely see radical changes. Although the airline business has always been more exposed than many other businesses to the economic cycles, the current crisis is likely to have more severe impacts. Moreover, even when the economies begin to show signs of a modest recovery, oil prices (that have a disproportional leverage on airlines) could go up again.

- Financing corporate debt is reported to be three times more expensive than it was in 2005, not to mention that the financial institutions are now less willing to extend credit.[6] Despite the higher cost of debt, many airlines are taking on debt to have liquidity during the downturn and, in some cases, to fund pension obligations.

- The role of governments is expected to be expanded worldwide, in areas ranging from transparency to protectionist policies that, in turn, will range from their impact on financial institutions to regulations to protect the environment. Airlines could easily see an increase in government intervention ranging from a greater focus on rules relating to common carriage, to the implementation of a global "carbon tax" (making fuel expensive again) to control emissions and protect the environment.

- Disposable income in developed regions, particularly the US, may not grow for a variety of reasons, such as the maturing of Baby Boomers and the impact of the collapse of the housing and the stock markets on the Boomers' retirement plans. Ageing populations will be less productive and produce higher cost in the healthcare sector. Also, over time, the cost of the present debt-financed stimulus packages will inevitably lead to higher taxes. Consequently, the growth in air travel in the developed regions is not likely to be at anywhere near the historical level of twice the rate of growth of the economies.

- The growth in income and consumption is expected to move to Asia, changing the economic power of global economies, and the varying potency of emerging markets. The expected strong economic growth of the new economies, such as those that will be discussed in Chapter 6, will lead to much higher growth rates in air travel. However, this higher growth rate, being applied to relatively small bases, may not offset the lower rates of growth in air travel in the developed regions, at least not in the near term.

- Consumer behavior, particularly relating to purchases, will change dramatically as a result of numerous forces. In the US, the older generation now admits to spending beyond its means, and the younger generation appears to be less addicted to "materialistic" lifestyles. The Millennial generation, for example, focuses more on shopping smartly; having smart mobility, as well as smart communications channels than on the fulfillment of the prestige needs. This generation will most certainly impact airline product decisions.

- Social technology is now commonly exploited by technology-savvy travelers to produce a major impact on the image of the airline industry, as well as individual airlines. Here are just two examples. In February 2008, passengers, stranded for about 10 hours on a jetBlue airplane during a snow and ice storm, were able to communicate with the media in real-time about their experience while on the aircraft. In July 2009, a musician became an Internet sensation by making a video (and placing it on YouTube) with a song on how United Airlines broke his guitar.

- There is also rapid growth in the usage of "virtual meeting technology," such as Cisco's "Telepresence."

Consider now some factors relating more directly to airlines that would have been unimaginable a few years ago.

- A significant percentage of consumers in the US would prefer to fly low-cost airlines, instead of full-service airlines because they perceive the product to be better.

- A small feeder airline in the US (Republic Airways) would pursue a bold strategy to acquire Frontier (a network airline based in Denver, Colorado) and Midwest (an ailing, full-service carrier based in Milwaukee, Wisconsin).

- Southwest would make an attempt to buy Frontier, an airline with a totally different fleet, or consider flying outside the US.

- Jet Airways and Kingfisher Airlines (two fierce competitors based in India) would agree to a strategic code-share agreement.

- Jetstar, a low-cost subsidiary of Qantas, would grow faster and be more profitable than its parent.
- Malév, Hungary's national flag carrier, would come, indirectly, to be managed by Aeroflot.
- Aer Lingus, the Dublin-based airline, would consider setting up operations at Dulles Airport in the US, to fly Dulles-Madrid with its own aircraft and crews, and with United managing the revenue.
- Passengers would pay as little as US$2,000 for a transatlantic Business Class ticket from top-brand airlines.
- Two US airlines (American and Delta) would battle over the purchase of part of Japan Airlines.
- The Japanese and Canadian governments would struggle to find different ways to bail out their heretofore flag carriers.
- The Japanese government would even consider discussing an Open Skies treaty with the US government, or that Haneda Airport would once again become an international airport (possibly with 24-hour operations).
- An Asian low-cost airline, Air Asia, would evolve into a global airline with hub connectivity, but without interlining.
- GOL, a successful low-cost Brazilian airline, would take over Brazil's long-ailing flag carrier, Varig, and survive.
- Lufthansa would have ownership in airlines based in Austria, Belgium, Italy, Switzerland, Turkey, the UK, and the US; and that it would continue to pursue equity ownership in airlines based in Poland and Scandinavia.
- A survey carried out by Ryanair (with over 120,000 passengers) would show that 42 percent of the passengers

would be willing to stand on short (one hour) flights if they could pay 50 percent lower than seated passengers.[7]

- Many low-cost airlines would add complexity to their original simple business models through the development of code-share agreements, the use of global distribution systems, and travel agents to distribute their seats.

- Airlines would generate up to a third of their total income from non-ticket revenues.

Clearly, all of the aforementioned points (relating directly or indirectly to airlines) do not apply to markets worldwide. First, the American notion of consumerism is quite different from many other parts of the world. For example, the "trading down" phenomenon is less evident in many parts of Latin America and Asia where, on the contrary, the aspiration is to consume premium products and services. Second, the fact that consumers entered the recession loaded with debt (and some US consumers may be paying this debt the rest of their lives) is also a mostly US phenomenon. Again, it is not the case in the rest of the world, and, as such, the consumers in those regions may not feel the need to "trade down." Third, the high-level penetration of Internet-related technology and its ramifications are not yet a totally global phenomena. For example, there are many countries in the world where most travelers still use the traditional brick-and-mortar travel agents and airline city ticket offices to make reservations and purchase tickets. Consequently, the new normal described in this section applies mostly to the US, parts of Europe, and parts of Asia.

Airlines have obviously been trying intensively to deal with the current chaos as best as they can, given their inherent constraints, such as government ownership and control rules, restrictive regulations contained in many Air Services Agreements, as well as political intervention. However, corporate leadership in the future (mostly in western economies and some parts of Asia) will need to deal with what some business analysts refer to as the "new normal,"[8] "permanent crisis,"[9] and the landscape of "inherent unpredictability."[10] The key point, however, is that managing within the new normal does not mean that the emerging business environment will simply be a huge challenge. There will also be enormous opportunities for those managements who are prepared to lead in the new environment using non-traditional business models, developed with best business practices and capabilities (as well as some imagination), and who work around the constraints internal to the airline industry.

There are, basically, three fundamental growth drivers in commercial aviation: growing world economies, enabling technology capability, and the removal of some limiting government-imposed regulatory constraints on airline commercial practices. Consider, first, economic growth. While the global air travel growth rate has varied from year to year over the past 30 years, it has correlated well with the growth in the global economies, resulting in an average rate of growth in traffic that has exceeded 5 percent per year. See Figure 1.5. As for the future, while the global GDP growth may only average around 3 percent per year for the next 20 years, the economies of certain countries (such as China and India in Asia and parts of Latin America) are expected to grow at much higher rates. In addition, setbacks in traffic growth in some regions could possibly be made up, to some extent, through the stimulation of traffic by growing airlines in other regions, such as from the operations of the airlines based in the Gulf region of the Middle East, and the lower- cost, lower-fare airlines starting operations in intercontinental markets (for example, AirAsia X and V Australia).

While the decline in premium travel is related to the state of the economy, it could improve to some extent as the economy improves, based on the experiences of the reduction in, and recovery of, premium travel during the years 2001 and 2003. In both cases, the traffic levels returned to their previous levels. However, even if the traffic levels were to return to their pre-economic crisis levels, the fare levels for travel in premium cabins in intercontinental markets are not likely to return to their previous levels for a couple of reasons. Some travelers previously choosing Premium Classes may switch from the higher-priced, flexible fares to the lower-priced, restricted fares. Next, some premium travelers buying Business Class and First Class tickets in intercontinental markets may switch to the Premium Economy cabins. Finally,

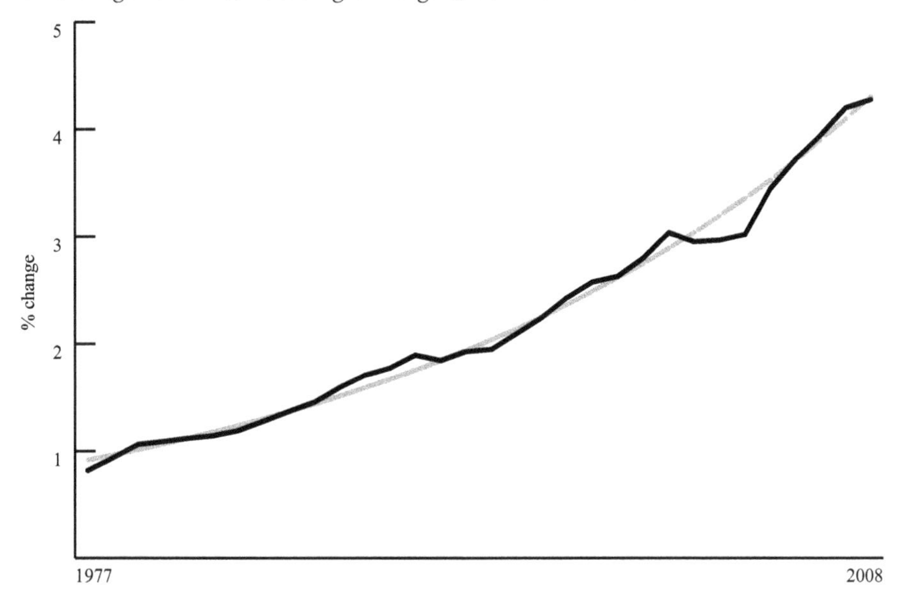

Figure 1.5 World Air Travel Growth Since 1977
Source: Boeing Airplanes, "Commercial Market Outlook," June 2009 and the ICAO.

some premium travelers may switch from the higher fares charged by traditional carriers to the premium services offered by the new intercontinental airlines, such as V Australia, or even low-fare subsidiaries of traditional network carriers, such as Jetstar. The second contributor to the growth of commercial aviation is the enabling technology capability. With respect to aircraft alone, each new generation has introduced technology that enabled airlines, in some combination, to reduce costs, reduce fares, or improve the product in terms of capacity, range, or frequency. This trend is expected to continue with the introduction of such airplanes as the Boeing 787, the Bombardier CSeries, and the Airbus A350. Imagine the number of markets that could be served nonstop from New York by the forthcoming Boeing 787 with a range of 8,000 miles. Suppose there is already a nonstop flight between JFK and Seoul, Korea. The 787 could enable additional nonstop service from Newark Airport. In a different case, the 787 could enable an airline to offer nonstop service between either JFK or Newark to Ahmadabad in India. Similarly, the forthcoming Bombardier CSeries could provide nonstop service to virtually any city in the

African continent from Lagos, Nigeria. In a different case, the CSeries could enable an airline to fly nonstop from New York to Cali, Colombia.

Other areas of technology (relating, directly and indirectly, to the Internet and mobile communications) will enable airlines to improve the experience of their passengers and the efficiency of their operations.[11] Examples include improvements in the capability of airlines to:

- engage with customers through the Internet to respond to their needs;
- increase ancillary revenue by implementing some elements of a merchandising model;
- recover mishandled baggage much more efficiently; and
- provide increasing facilities for self service.

The last example of technology capability provides more control to customers and saves money for airlines.

The third contributor to the potential growth of, and the sustainability of, reasonable profit margins in commercial aviation is the increase in commercial freedom for airlines. Elimination of regulatory constraints, enabling airlines to deal with "doing-business" issues, would lead to greater efficiency and optimal organizational structures. The three key areas in which government regulatory barriers need to be lowered relate to (a) market access, (b) airline ownership and control, and (c) the installation of efficient and effective new ATC systems. However, based on historical experience of airlines going through various crises, it is difficult to assume that governments will ease the really constraining regulations, allowing airlines to operate more like other businesses. On the contrary, it is possible that some governments may, in fact, become even more protective of their national carriers, preventing the winners from winning and the losers from losing. It is ironical that the exit barriers in the global airline industry are becoming higher than the entry barriers.

In light of the expected changes in the performance of global economies, the developments in the enabling technology capability and the potential prolongation of regulatory controls, airline leadership must look for innovative ways to transform their businesses to respond to the aforementioned new normal. The transformation of airline business models can range from simply improving products at one end of the spectrum, to implementing strategies that shift various degrees of risk to other members in the air travel supply chain in the middle of the spectrum, to distributing the product in radically different ways to meet the ultimate need of travelers, namely, personalized mobility, at the other end of the spectrum. This full-spectrum transformation of airline business models must be analyzed with a new mindset, based on the best practices from other businesses, around the numerous airline industry-specific constraints, and with some imagination.

Takeaways

Most airlines have done reasonably well in reducing their costs and capacity, enhancing their ancillary revenue, and increasing their liquidity to survive. However, despite the implementation of these strategies, a few carriers will not survive the current chaos and will lose their identity via consolidation. They are looking at crisis-driven change as a pain, as opposed to a gain. Furthermore, some are cutting their costs in areas that are weakening their core business, leaving their best markets and best customers unprotected. On the other hand, this is the best time to create a change (a corporate renewal), both within the industry as well as within the company. Resistance to change is likely to be less during industry crisis. Some carriers are taking an offensive strategy to exploit the new world, while others are taking a defensive strategy by slimming down and hoarding cash to survive in the near term. Outside the airline industry, one can recall what Lou Gerstner did with IBM in the 1990s when he gave up the hardware business and went after the software and the services business. The new normal will be a time of change, structurally and fundamentally. There will be a shift in the power centers of the global aviation market, based on, for example, the expansion of Gulf-based carriers during the past five years and this downturn. The inability to make tough decisions in a timely manner can destroy individual companies, as well as an industry. One only needs to look at the experience of the US auto industry, the subject of the next chapter.

Notes

1 "Deeper Losses Forecast—Falling Yields, Rising Fuel Costs," Press Release No. 37, International Air Transport Association Corporate Communications. Department, Geneva, Switzerland, 15 September, 2009.
2 "Premium Traffic Monitor," *IATA Economics*, June, 2009, p .1.
3 Ballantyne, Tom, "Boom Time for Budget Carriers," *Orient Aviation*, September 2009, p. 14.
4 Sobie, Brendan, "Competition in Brazil Heats up," *Airline Business*, September 2009, p. 16.
5 Ormes, Ian, "Supply & Demand," *Airlines International*, June-AGM 2009, pp. 40–42.
6 Charan, Ram, "My (Recovery) Playbook," *Fortune*, 31, August 2009, p. 61.
7 Reported in *Air Transport News*, 22 July, 2009, by Ryanair.
8 Davis, Ian, "The New Normal," *McKinsey Quarterly*, Number 3, 2009, pp. 26–28 and Geoff Colvin, *The Upside of the Downturn: Ten Management Strategies to Prevail in the Recession and Thrive in the Aftermath* (New York, NY: Penguin Group, 2009), Chapter 2.
9 Heifetz, Ronald, Grashow, Alexander, and Marty Linsky, "Leadership in a (Permanent) Crisis," *Harvard Business Review*, July-August 2009, pp. 62–9.
10 Ramo, Joshua Cooper, *The Age of the Unthinkable* (New York, NY: Little, Brown and Company, 2009), front jacket flap.
11 Peters, Jim, "Ten Technology Advances that Will Change Air Travel," *A New Frontiers Paper*, SITA, 2009.

Chapter 2

Learning from Other Struggles: The Auto Industry

Although there are clear differences between the automobile and the airline industries, there are also many similarities. The following are just ten of the many similarities.

1. Irrelevant and broken business models (reliance, until recently, on a single approach to manufacturing/operations and selling).

2. Ability of lower cost and higher product quality competitors to make enormous incursions in the market share.

3. Powerful influence of labor (both with respect to costs as well as productivity).

4. Management slow in adapting to the realities of the changing and evolving markets.

5. Potential opportunities for game changers, or format invaders through increasing levels of productivity and capability.

6. High profile industries with significant political interference.

7. Challenge to solve problems relating to the climate change.

8. Lack of comprehensive "due diligence" in M&A.

9. Overcapacity at the present time.

10. Enormous growth potential from the desire for personal mobility in emerging markets.

In light of the dramatic problems that have been facing the incumbent automobile companies, mostly, the Detroit Three, are there any insights for the airline industry? This chapter provides a background to the automobile industry consisting of the focused competitors including alliances, the unfocused competitors including a case study on DaimlerChrysler, and what is next for the automakers. Of course, some insights for airlines based on the experience of the automakers, both successful and unsuccessful producers, are also presented. Even more important, given the similarities between the two industries, there are high growth opportunities globally for both industries.

The Focused Competitors

Toyota

While many write-ups on Toyota focus upon its success in areas such as lean manufacturing and just-in-time production, this discussion will focus on the more difficult to copy aspects such as the company's culture. There are dozens of excellent books that have been written on the critical success factors of Toyota. The key, of course, is the consistent quality of the products that, in the case of Toyota, could be considered to have been started with the Corolla (one of the best selling cars in the world), and followed by the Camry. Then, having solidified its dominance in family cars, Toyota moved into the higher-margin, luxury segment with its Lexus offering, and now has expanded to capture the younger segment, specifically Generation Y or the Millennial generation, with its Scion (for more on the Millennials, see Chapter 3).

How does Toyota capture these markets? First, Toyota fosters a culture of experimentation. The company rewards the knowledge and experience gained from experimentation, rather than harping on success or failure. Experimentation promotes trying new things in an effort to stay ahead of the times. These practices are not only a motivating force, but a responsibility of employees that have abled the company to be of service to people. An important driver in the culture of the company is challenging the status quo and discovering something better, and therefore the company makes significant time and financial investments in order to allow experimentation. One example is its early investment in hybrid vehicles at a time when Toyota's competitors considered hybrid technology to be unproven and, as such, were unsure of the vehicle's appeal to consumers. For Toyota, even partial information gained through experimentation is an opportunity as it can be a milestone on the path to achieving a possible solution. And it did. Experimentation led to the development of the popular hybrid car, the Toyota Prius. One of the most remarkable things about the Prius is that Honda developed a hybrid at the same time as Toyota, yet Toyota made a marketing success out of their product, leading to the present situation where Toyota effectively owns the hybrid market. Similar experiments were made during the introduction of the Scion, each time conducting market research to ascertain information such as consumer awareness of the product, sales and delivery processes, and consumers' experiences with respect to the whole purchase process as well as ownership of the vehicle. An important means of promoting this culture of experimentation was through multi-stage training programs focusing on multi-step problem solving skills.

A second attribute of Toyota's culture includes successfully anticipating customer wants and needs, which allowed the company to penetrate key market segments in the US, such as Lexus in the luxury market and Scion in the youth market. A third attribute is a high level of customer service. Lexus has been a leader in the auto industry with respect to customer service. Specifically, Lexus was one of the first to offer such services as a loaner vehicle of a comparable class while a customer's vehicle was being serviced, 24-hour emergency roadside assistance and customer shuttle service, and complimentary gas and car washes. Customers could even view service work being done on their vehicle while they waited. Additionally, some dealers provided complimentary amenities or courtesies, including providing meal coupons to customers waiting for their cars to be serviced. One way in which Toyota has instilled a sense of top notch customer service into Lexus dealers was by having them stay at the Ritz-Carlton Hotel in Osaka. The idea was for the dealers to get a feel for the level of service that they, in turn, would be expected to offer their customers (for more on Customer Experience and the Ritz-Carlton, see Chapter 3). Also, management attended the Fuji Lexus College near Tokyo to learn about the hospitality philosophy, returning twice a year to review and to share best practices with others.

Finally, Toyota has mastered the concept of being "glocal" (being global while at the same time catering to local tastes). Toyota began as a small, local company. Today the company conducts its worldwide business from the same location, despite its global presence. When the company operates in other countries, it is truly sensitive to the needs and requirements of those countries. Management has tried to become extremely "local-friendly," recognizing that the company's operations can significantly affect the local economy and its people. Toyota acknowledges that there are definite challenges to local customization. For example, one challenge is to figure out how to scale the design of a vehicle and its manufacture to provide cost benefits, and at the same time satisfy local needs for customization. Toyota has been able to do both without compromising either. However, Toyota also sees the benefits derived from local customization. Local customization also exposes Toyota to conditions that do not exist in its native country, and these differing conditions help the company to conduct research and development. For example, Toyota was able to implement innovations to meet the unique requirements with respect to the Middle East's climate after having the opportunity to drive across the Arabian Peninsula in the 1970s. Specifically, the company's engineers were able to conduct tests in extreme temperatures, leading to the development of steering wheels and dashboards that could withstand such extreme temperatures as well as gauges that could accurately measure them.[1]

Porsche

Porsche was close to bankruptcy in the early 1990s, when the company took a huge hit due to the state of the US economy and the stock market crash of 1987. Sales volume plummeted from 50,000 vehicles in 1986 to 14,000 in 1993. The company also suffered from a stagnant product line as well as leadership issues.[2] However, in 2003, a decade after Porsche appointed (now former) CEO Wendelin Wiedeking, it became the most profitable automaker in the world, even outperforming the Japanese. The company did a total turnaround from the days where it was plagued with declining sales, losses, and liquidity issues.[3] How did the company make this dramatic turnaround and reinvent itself? First, two ex-Toyota managers were brought into Porsche, and they completely turned the production and design process around in Stuttgart. This transformation indicates that even a prestigious German car manufacturer has to adapt to the state-of-the-industry processes. Second, it launched a new vehicle, the Porsche Boxster. Third, the company made the decision to expand into the sport utility market segment of the auto industry, with the newly developed SUV, the Cayenne. This was a controversial and risky decision, as many were skeptical and believed that moving into the SUV market segment would dilute the brand. However, Wiedeking recognized the need to not lose sight of the brand's image by insisting that the new vehicle had to be "a real Porsche in terms of chassis, performance, and design,"[4] all key strengths of the brand. Furthermore, a survey indicated that a typical customer of the company was reported to already possess a limo or sedan, a sports car, and an SUV, and many of the company's customers showed an interest in an SUV offering by Porsche. Finally, the Cayenne allowed Porsche to penetrate high growth, developing markets more successfully, especially where the quality of the roads do not allow one to own a 911. On the same note, other markets were characterized by business leaders who do not drive themselves, but still wished to have a Porsche. Therefore, the Cayenne fit both of these markets beautifully, and continues to sell very well.

One unusual characteristic of the production of Porsche's SUV was the decision on the location of its production. While Alfa Romeo, Aston Martin, Ferrari, and Lamborghini are all built in their historic countries of origin, in contrast, Porsche has outsourced production of its Boxster and Cayenne models to Valmet in Finland (only the engines are built in Stuttgart), and is switching the Valmet contract to another outsourced producer, Magna, in

Austria. The production of the Cayenne's body is outsourced to VW, who build it in Bratislava, Slovakia, and then it is shipped to Germany for completion. It is assembled in Leipzig, at a low-cost factory in former East Germany. Part of the genius of Porsche is its ability to manage the "built in Germany" image, even though the vehicles are only partially built in Germany. Porsche has managed to maintain its elite product image in spite of having the highest outsourcing level of all major car manufacturers. For example, the 911 series has only 20 percent Porsche content, the Cayenne just 10 percent.

Another genius of Porsche is that the company has a flexible, outsourced, production system, which transfers a lot of volume risk onto the outsourced suppliers. This is also a different path than competitors BMW and Mercedes-Benz, who have set up their SUV production plants in the southern states of the US (BMW in Spartanburg, South Carolina and Mercedes-Benz in Tuscaloosa, Alabama) in an effort to be closer to the US market. Porsche's success illustrates that even a luxury product can be built with a lot of global sourcing, if the quality is rigidly controlled.

Another unusual characteristic implemented by Porsche is its unique combination of applying Japanese production processes of lean production while still maintaining the famous German reputation of craftsmanship and quality. Specifically, in the early 1990s, Wiedeking toured production sites of automakers including Honda, Nissan, and Toyota, with Porsche management in an effort to illustrate the vast differences in their lean production processes versus those wasteful ones of Porsche.[5]

Finally, with the possible end of the SUV era in sight, Porsche has introduced a new product, the Panamera: a four-door sedan that is expected to compete with the Aston Martin Rapide and the Mercedes-Benz CLS. Also, Porsche quickly introduced a powerful diesel engine for its Cayenne SUV to cope with rising fuel prices that started to threaten the success of the SUVs, in the US and in Europe. The insight here is that imminent changes in consumer behavior require constant adaptations of the product line.

Alliances

Unlike so many consolidations in the auto industry, the Renault-Nissan strategic alliance has been extremely successful. What was their secret? First, the two companies leveraged their alliance to produce the greatest economies of scale. For example, Renault was able to take advantage of open capacity at Nissan's production plant in Aguascalientes, Mexico in order to produce its Renault Clio for the Latin American market. This open capacity allowed Renault to jump into production quickly, and with far less investment cost. While this aspect seems like the main basic premise for engaging in consolidation, many companies seem to not have been able to achieve this goal. Examples include BMW and Rover, GM and Isuzu, and Hyundai and Kia. Why?

It appears that in many situations, the "merged" companies actually remain separate in areas ranging from design and engineering to production. Renault and Nissan, on the other hand, worked as one unit from the very beginning. While Carlos Ghosn's immediate actions were to rescue Nissan, even at that point synergies were being achieved, such as cost savings realized from the consolidation of purchasing. In the case of the plant in Aguascalientes, Mexico, there were also commonalities between Renault's Clio and Nissan's Platina which eased some production challenges. As Renault and Nissan continue to commonize the architecture and production of vehicles and powertrains, they could become even more efficient as they achieve the ability to produce a variety of vehicles under both brand names at any given facility. Second, the alliance has been successful because the management of the joint operation understands that consumers desire both a high level of quality as well as unique styling. In most cases, though, it does not matter if the brands or models are produced together as long as these attributes are maintained.[6]

The joint venture between India's Maruti and Japan's Suzuki has been a long-term profitable success. Initially, the venture was a government project in the 1980s with the goal of introducing a low-cost modern automobile targeting the Indian middle-class consumer. However, in 2007–2008, Maruti-Suzuki was still a dominant player in India, capturing over 50 percent market share. Specifically, foreign automaker, Suzuki, a small-car specialist joined forces with local automaker Maruti. The current lineup is comprised of almost all Suzukis, including the Alto and the higher-end Swift. While the Japanese possess more equity and are the drivers in this joint operation, many Indians still perceive the venture as a local company.[7]

Renault turned the Romanian automaker, Dacia around after it initially acquired a controlling stake in the flailing and previously state-owned Dacia in 1999. How did they do it? Renault-Dacia introduced the Logan: a low-cost car for emerging markets in 2004. While its main plant is located outside of Bucharest, the company has seven other assembly plants in other locations including Brazil, Colombia, India, two in Iran, Morocco, and Russia. The car has taken off, with the millionth Logan produced in mid-2008. The car has become one of the French company's most profitable offerings. Specifically, Renault set out to introduce an "affordable" car, meaning one that could be offered for US$6,000. The company thought that it would be able to attain this goal by taking almost all pieces of the car from previous or existing models. The automaker was able to keep costs down by keeping the car simple, with fewer parts than the high end cars and less electronic devices. For example, the Logan's dashboard is a one piece molding. Moreover, many of the machines at the plant outside of Bucharest are actually refurbished from the company's operations in France. One report indicates that the Logan production costs were lower than half of other compact car offerings in Europe. Besides the lower price, lower maintenance costs represent another value-adding feature. Although the car was designed to be offered at US$6,000, it has been reported that the Logan is actually selling for closer to US$9,000, and consumers are even paying US$13,000 for features including air conditioning and power windows.[8] As in the case of Toyota, Logan's design was customized for local conditions, for example, special chassis for poor road conditions, and a special engine to handle lower quality fuel.

The synergy potential of alliances goes far beyond joint marketing. To exploit it on the development and production side, partners have to agree to common standards of platforms, joint sourcing, and an integrated production design. All this requires giving up a lot of autonomy, which makes it a difficult task to achieve.

Insights for Airlines

- The key drivers in the culture of Toyota are: challenging the status quo and discovering something better; making significant time and financial investments in order to allow experimentation; implementing comprehensive training programs and problem solving skills; and developing top-notch customer programs. In the airline industry, a few airlines have attempted to introduce a similar culture in their own management environment. With respect to experimentation, examples include Continental leading the way to the utilization of a narrow-body aircraft (the Boeing 757) in intercontinental markets and Air Canada leading the way to unbundling products and à la carte pricing.

- Porsche transformed itself by hiring Toyota managers to overhaul its production and design process, while maintaining the German reputation of craftsmanship and quality. They launched new vehicles, the Boxster and the controversial Cayenne, its SUV, and transferred a lot of volume risk onto the outsourced suppliers. In the case of airlines, a number of airlines, such as BA, Continental, Delta, KLM, SAS, and United, also experimented by introducing their own low-cost, low-fare subsidiaries, but they did not succeed. Qantas, on the other hand, has been successful with its initiative, Jetstar, with units based in Australia and in Asia. The critical success factors lie in the detail of the design and its implementation without compromising the brand.

- The key driver of success of both the Renault-Nissan and the Renault-Dacia strategic alliances was the operation as one company to achieve economies of scale and integrated product designs. In the airline industry, alliances have had limited success due, in part, to the existence of government regulations and, in part, to the unwillingness of some partners to give up their autonomy. Compare this last point with the benefit achieved by Dacia to produce the Logan, a successful low-cost car for emerging markets. The initial success stories in the airline industry are the strategic alliance implemented between Northwest and KLM, with truly integrated hubs at Narita, Detroit, and Amsterdam, and the partnership within the LAN group, which represents one brand with four production units.

The Unfocused Competitors

The Detroit Three

A mere 40 years ago, the Detroit Three automotive manufacturers dominated the American marketplace with their vehicles accounting for more than 9 of 10 automobile sales in the US.[9] Some may remember the famous sayings, "What's good for General Motors is good for the country" and "Doing it the GM way." However, this is certainly no longer the case. At the end of 2008, GM and Chrysler needed the US government to bail them out and keep their companies running. Chrysler filed for bankruptcy at the end of April 2009 and General Motors filed at the beginning of June 2009. What happened to these American icons?

The downfall could be traced back to as early as the 1960s, when Volkswagen started offering its products, particularly the Beetle, in the US. The Japanese emerged as players in the small car market segment in the 1970s and the trend continued in the 1980s as they moved into the mid-sized sedan market that was traditionally the strength of the Detroit Three. Also in the 1980s was the emergence of foreign competition in the luxury vehicle market. Lincoln and Cadillac used to be leaders in the highly profitable American luxury vehicle market, but have not been so since the mid-1980s. The luxury market segment has been taken over by numerous foreign brands, such as BMW, Mercedes-Benz, Infiniti, and Lexus.

There was a glimpse of hope for the Detroit Three in the 1990s with the sport utility vehicle (SUV) boom. Specifically, Chrysler,

Ford, and GM dominated the marketplace with approximately 90 percent of SUV sales at the time. However, foreign automakers were quick to enter the SUV game. By 2003, more than a dozen different SUV models were offered by foreign automakers. There was also some hope for the American car makers during the 1990s through minivan sales. Chrysler, in particular, excelled in this market segment. However, in the late 1990s, the Japanese totally changed the landscape of this segment with Honda's introduction of the Odyssey minivan, which took off due to Honda's reputation for quality. Other foreign automakers such as Toyota and Nissan also began to introduce minivan offerings, further cutting into the American market share of this segment. The last arena where the Detroit Three still held their ground was the pickup truck market. Ford, in particular, excelled in this market segment. The Ford F-series had a long running sales record—one of the best-selling vehicles in the US since the late 1970s. However, in the late 1990s, the Japanese also changed the landscape of this market segment with the introduction of the Toyota Tundra, which was reputed to be a leader in quality. What went wrong with the Detroit Three? Here are nine possible contributing causes.

First, the Detroit Three were trying to be all things to all people. For example, Alfred Sloan, one of GM's great early leaders, proclaimed that GM could offer "a car for every purse and purpose."[10] Such a strategy can be viable, as demonstrated by Toyota and the VW Group, if the manufacturer is (a) smart in using platform strategies that produce economies of scale throughout the production line, and (b) the manufacturer capitalizes on the synergies in distribution.

Second, tied to the problem of trying to be everything to everyone was the Detroit Three's need to be the biggest. For decades, they focused on market share and volume rather than steady profitability. This emphasis on size made them the victim during economic cycles by needing to make enough profits in the good times to compensate for losses when the economy was in a downturn.

Third, the Detroit Three have historically had too many dealerships (the 2009 bankruptcy proceedings of both Chrysler and GM allowed them to reduce the number of dealerships). Japanese competitors Honda and Toyota have not been burdened by the enormous networks of dealers that the Detroit Three have created. Historically, the Japanese have had a fraction of the number of dealerships that the Detroit Three have had to sell the same number of vehicles. Such an efficient network allows for the dealers to communicate with customers and management much more easily, and also to have a larger profit margin on each vehicle sold.

Fourth, the Detroit Three did not understand and anticipate their customers' needs. They were complacent and less innovative than their competitors, and their product development cycles were too long. They were also too slow to adopt lean manufacturing techniques. Foreign automakers reacted faster to changing consumer tastes as well as their own needs. As a result of building to inventory rather than to dealer/customer orders, Chrysler, in particular, was saddled with huge finished goods inventories during the down cycles. All of the Detroit Three burned through billions of dollars in rebates and dealer allowances in order to move unwanted product.

Fifth, the Detroit Three shifted their focus from product to financing, specifically by focusing on incentives, such as zero percent financing, no money down, etcetera, rather than on building profitable products at prices consumers will accept. They began to focus on short-term volume over the long-term elements of credibility and product enhancement. The "rebate strategy" actually has had greater cumulative financial impact than even the benefit issues (addressed below). Due to the fact that they produced so much inventory that consumers did not want at prices that would produce profits (systemic overcapacity of 20 percent or more), and because the quality of the products they built led to uncompetitive residual values, they resorted to a hapless strategy of rebates. The result was a classic vicious cycle. So-called "marketing costs" reached about US$4000 per unit sold. Rebates did not create more demand, they just moved it up from the next quarter, and lower effective selling prices eroded resale value, so even "loyal" consumers needed rebates if they wanted to trade in or trade up to new cars. The point simply is that the Detroit Three altogether shifted their attention to the more lucrative financial arms of their businesses rather than focusing on fixing their core businesses.

Sixth, the Detroit Three mismanaged their labor relations. The automakers have tied their own hands in numerous ways due to unsustainable agreements with the United Auto Workers (UAW). For example, the jobs bank agreements signed with the UAW in the mid-1980s limited their flexibility in ways such as the ability to right-size production because they had to continue to pay UAW workers (including benefits) whether they were working or not. Management needed to increase productivity through an increase in automation as well as more flexible manufacturing processes while the union wanted to protect current workers, avoid layoffs, and discourage outsourcing. The agreement led to an imbalance in supply and demand. The automakers produced vehicles when there was little or no demand for them just to keep workers working, but then had to turn around and offer huge incentives to buyers. This program became a major source of competitive disadvantage for the Detroit Three compared to their competition in Europe and Asia.[11]

Furthermore, there is the enormous burden of the healthcare and pension expenses for retirees of the Detroit Three (in excess of US$1500 per vehicle), a cost that their foreign competitors do not incur. Patrick Gaughan notes that this financial obligation goes much further than just the retirees of the Detroit Three; the Detroit Three have never really been able to sever their ties with the suppliers they once owned, in terms of their workers, due to the UAW. For example, GM separated from its supplier, Delphi, in the late 1990s, but later, when Delphi filed for bankruptcy, GM had to honor its obligations to Delphi employees. Ford had a similar situation occur with its supplier, Visteon. Years after it had separated from the company it was forced to rehire many of Visteon's employees.[12]

And in terms of benefits, there has also been the VEBA issue. VEBA, which stands for Voluntary Employee Beneficial Association, was a transition vehicle to move the employer-financed health benefit plans to the union. The funds that were in the plan were literally transferred from the company to the union, at which time the union assumed both administrative and financial responsibility. While this seemed to be a positive move for the Detroit Three in terms of decreasing their liabilities, the VEBAs had problems—one of which was that the VEBAs were severely underfunded. The VEBAs did poorly, presumably because these transitional vehicles were supposed to be just that—transitional—and that the ultimate goal was to get workers to purchase an individual insurance health plan.[13]

The agreements with the UAW have held the Detroit Three back from achieving the efficiency enjoyed by the foreign automakers operating plants in the US. Micheline Maynard emphasizes that the actual obstacle is the UAW's master agreement with each company, including the staffing requirements, the holiday requirements, and all of the benefits including pension pay. It is important to note that in addition to the master agreement, there are also additional individual contracts for each plant. All of these requirements detailed in these documents produce additional cost and complexity which the foreign automakers operating plants in the US do not incur. One example that illustrates this point was reported by an individual who worked at one of the Detroit Three, and then went to a foreign automaker operating in the US. A GM contract with the UAW required three workers on an assembly line at GM, whereas in the same situation at Toyota, there was only one worker.[14] However, as Maynard notes, there are in fact two signatures on each contract; one by the UAW and one by the US automaker. Consequently, the UAW, alone, cannot be blamed for imposing the requirements when the automakers negotiated and agreed to all of them.

The seventh contributing cause is that, starting in the mid-1980s and following the disastrous "labor peace at any price" jobs bank agreements with the UAW, in which the strategy was, above all, to keep the plants in production because they were obliged to pay the UAW workers whether or not they were building cars, the automakers decided to either buy car rental companies or find a way to subsidize them. The buy-back programs and other allowances were production-driven, not market-driven. In the same era, new car costs began to rise rapidly, and there was a demand for more affordable "near new" vehicles. Dealers were then conflicted, because although they had profitable used-car sales, they were more challenged to increase new-car business.

There was clearly systemic overcapacity, unclear accounting (see note below on this), and lack of profitability in the operation of most of these US rental car companies. Operating overcapacity was the direct result of manufacturing overcapacity. Like the manufacturers, the rental car companies (or subsidiaries) overlooked operating profit in favor of market share and per-car allowances. The more vehicles churned through the fleets, the more dollars were given from the factory. Hertz, the strongest brand, remained profitable, while Dollar Thrifty, the smallest contender, has simply hung on. Is it a coincidence that Enterprise, a family-controlled, entrepreneurial business with no entanglements or ownership interest from the Detroit Three is now the world's largest car rental company? Undoubtedly not.

Another negative factor relates to the auto brands. Since rental fleet operators were the customers of last resort and suppliers of more affordable "near new" vehicles, they were generally consigned the least desirable models, vehicles whose production cycles were extended on the basis of artificial demand. This, too, diverted energy and attention from the more rapid development of more desirable models. Proof of this principle is evident in the very different fleet management policies implemented by car rental operators once they became standalone companies with, or consisting of, more diverse inventory for both rental and resale, smaller fleets (less overcapacity) and higher pricing, including unbundled pricing (addition of various fees for use of airport facilities, registration fees, refueling, etcetera), and addition of more interesting special vehicles (hybrids, sports cars, import brands).

In terms of accounting, as referred to above, it is important to note that one of the largest costs on a rental car company's P&L statement is depreciation on their large, and highly leveraged, fixed asset base, their fleet. Depreciation is, of course, only an accounting estimate. The real value is only established at liquidation. Unlike some aircraft, mass market cars never appreciate, and there can be large discrepancies between estimates on the books and actual inventory value. Thus, when United acquired Hertz, at one stroke it became the world's largest used-car reseller. Not long after, they sold the company.

Eighth, the Detroit Three spent more time lobbying for government protection from fuel efficiency requirements (and exploiting loopholes for pickups and other light-weight trucks) than improving their products. Corporate Average Fuel Economy

(CAFE) requirements were first introduced in 1975 in an effort to decrease the consumption of energy by increasing the fuel economy of cars and light trucks.[15] In December 2007, the US President authorized a bill that ultimately requires automakers to achieve a 35-mpg fuel economy standard by the year 2020. Moreover, the new system also changes the way in which CAFE requirements are administered; the 35-mpg standard requirement for 2020 will apply to the entire industry, regardless of whether the vehicle is a car, a pickup truck, an SUV. In contrast, the current system requires 27.5 mpg for passenger cars and 22.2 mpg for light trucks, including minivans, SUVs, and pickups, but the requirements are calculated as an average over a car manufacturer's entire fleet of offerings. This is where the US automakers are particularly at risk with respect to the new system as Honda and Toyota have relatively less offerings in the pickup and full-size SUV segments, and therefore have a better chance at exceeding the standard. The American automakers' lineups, which are currently heavily weighted toward these segments, could have a difficult time meeting the standards under the new system.[16]

Ninth, for the most part, the Detroit Three did not succeed in their M&A and strategic alliance activities. Among the many strategic mistakes in this area, the DaimlerChrysler merger (discussed below) is worthy of study as it provides significant insights for both automobile and airline managements. The marriage was supposed to be a win-win situation for both automakers. However, in reality, Daimler paid for Chrysler in the beginning and then again in the end for Cerberus to take Chrysler. Meanwhile, Chrysler ended up in worse shape due to the merger, and ended up going into bankruptcy again.

Why has Detroit failed? A combination of the above nine contributors can be summarized into: denial; lack of focus; unwillingness to adapt; inability to make difficult decisions; detrimental effects of size and inertia; and absence of consumer insight and market foresight.[17] Only their incredible size and inertia enabled them to survive as long as they have, but even this conveyed a false sense of competence and an inappropriate and self-serving confirmation of failed strategies. The foreign automakers producing in the US ("transplants") became the format invaders in the auto industry. Initially, the auto industry proved to be too large and too unforgiving for startups to have a really significant impact. However, the situation changed with the successful invasion of US manufacturing territory by nimble competitors with a better production model (lean and flexible manufacturing and a partnering approach with quality suppliers), superior employee relations (non-union environment and better training), a more efficient distribution strategy (fewer and better dealers), and faster cycling time. The Detroit Three did attempt to fight back. For example, GM created Saturn, which was not only as close to a "transplant" model as GM could get, but which also re-defined the car buying experience in the US. But, even that eventually evolved into a more conventional model. Saturn is now in the process of being phased out by GM after a potential deal to sell off the brand fell through. Hummer may be in the hands of a Chinese maker of heavy-duty trucks, and Saab is in the hands of a small sports car company.[18] On 10 July 2009, GM came out of bankruptcy and announced that "Business as usual is over at GM."[19]

Insights for Airlines

There are a number of parallels in the strategies followed by the Detroit Three and the airline industry. In the case of some US airlines, examples include:

- Focusing on market share rather than profitability.
- Trying to be all things to all people.
- Continuing to deny the enormous growth opportunities for existing and new low-cost carriers.
- Continuing to believe in the conventional business model (with only the need for minor adjustments, instead of the need for faster adoption of new and different business models).

- Focusing more on ancillary sources of revenue than on overcapacity, unproductive labor agreements, and inflexible production systems.

- Implementing poorly developed initiatives (Song, Ted, etc) to compete with LCCs (in the case of autos, Saturn was a much better initiative of GM just as Jetstar appears to be for Qantas).

- Becoming too dependent on corporate accounts and wholesalers.

- Using overcapacity as a defensive move (responding to competitive threats by adding more seats at low fares until the new entrant gives up or runs out of cash).

- Having a lower level of focus on product quality and reliability than their foreign counterparts, focusing on the higher margin products (premium cabins in intercontinental markets).

- Uncompetitive Business Class product in intercontinental markets.

- Insufficient focus on new, strong international competitors (for example, the Middle East Gulf-based airlines).

- Not paying sufficient attention to the potential government regulations regarding the protection of the environment.

Although there are many similarities in the structure, operations, and strategies implemented by the US auto industry and the US airlines, there are also significant differences that are not likely to lead the US airlines to the US Federal Government to be bailed out after the banks and the automakers. One major difference is that, unlike the foreign automakers that could set up plants in the US, foreign airlines are not authorized to operate within the US domestic markets. Virgin America tried. Although it did succeed in starting its operations (despite the severe restrictions placed on it), it has a very limited network and fleet to make a significant impact on the incumbent airlines (unlike the case of auto "transplants" that received incentives from communities, instead of restrictions). However, as discussed in an upcoming section about the possibility of the emergence of game-changing automakers, there could also be game-changing airlines based in the US.

DaimlerChrysler Merger Case Study

In analyzing the works of nearly a dozen authors on the subject, several themes surrounding the DaimlerChrysler merger have become quite clear.[20] First, it is interesting to note that both companies were actually profitable before entering the merger in 1998. Each entered the merger from a position of strength, which is typically not the case in a merger situation (consider, for example, BMW's purchase of Rover and Ford's acquisition of Jaguar around that time). Although Chrysler had bankruptcy scares in the past, it was nevertheless profitable at the time of the merger, and while it was the smallest of the Detroit Three automotive giants, the company had a strong presence in two products, the minivan and sport utility arenas. Daimler, despite its past of unsuccessful acquisitions and mergers, was not only one of the largest industrial companies in Europe (including aerospace and automotive divisions in its portfolio), but Daimler-Benz also was very strong in the luxury car arena.

From research, it is evident that there were several rationales for entering into the merger. On the American side, Chrysler was mainly a national company since almost its entire market share was contained in North America. Chrysler needed a means of tapping into the European and Asian marketplaces. Daimler-Benz would provide a gateway to the European network. Furthermore, Chrysler was facing direct competition from products being offered by the newly-established foreign manufacturing plants on US soil; specifically, both BMW and Mercedes had already opened their first assembly plants in the South. These companies had the advantage of cheaper labor through a non-unionized workforce as well as tax breaks offered by the state governments.

On the German side, it would appear to be difficult for Mercedes-Benz on its own to generate sufficient growth; Chrysler would provide a lower-end offering to augment the luxury brand. It has been reported that Daimler had reached the conclusion that in order to be a successful, global, automotive company, it needed to possess a full array of product offerings. The company needed a partner to help achieve this goal. If Daimler did not become proactive, it is conceivable that it could, itself, be bought, as had happened to two other luxury brands around that time, Jaguar and Rolls Royce (even though they were much smaller). Furthermore, Daimler had learned the hard way through its difficulties with the introduction of its A-Class car that it is not easy to enter a new market. The A-Class car was launched in 1997 to help the brand compete with small car offerings from brands such as VW. Some analysts have stated that Mercedes hoped that its brand name would transfer to this totally new offering, a new product in a new class of the marketplace. It was a success at first, until the new A-Class product was literally overturned in a test drive. Negative publicity mounted over the fact that the car had flipped. Analysis of the situation eventually led to the introduction of the Electronic Stabilization Program (ESP), but negative publicity continued. Finally, it was reported that by combining such functions as purchasing and research and development, as well as information systems, a several billion dollars savings could be yielded annually. It appeared that there were so many synergies to be enjoyed through merging. Consequently, at the apparently shallow level of analysis, it seemed like a perfect match.

In reality, the so-called match would appear to be a disaster. Several writers have synthesized key reasons as to why the merger was a failure. First, the merger was supposed to be a merger of equals, but in reality it was Daimler acquiring Chrysler. Daimler was always in the driver's seat. It was reported that in conjunction with this point, the Germans had developed a "game plan" in terms of the merger, while the Americans had none. Also in conjunction with the original point is the fact that HR from either Daimler or Chrysler had almost no role in the merger. The merger was presumably conducted by senior management without much input from the HR groups. It has been reported that the merger meetings focused upon the "hard" issues, such as financial and legal matters, rather than the "soft" issues of culture and people. Specifically, it has been stated that managements were aware that there would be cultural issues, but yet made little attempt to address them. The cultural mixing of the two companies in terms of the merger has been referred to as like that of "oil and water" for several reasons. The companies did not operate in the same manner. The Americans were used to being volume producers whereas the Germans were luxury car manufacturers (mass production versus customization). The Americans were more casual in both their demeanor and in their way of conducting business, whereas the Germans were formal and highly structured. There were language, salary, and time zone difference issues. Specifically, nearly none of the American executives spoke German, while almost all of the German executives spoke English. Chrysler gave its executives larger salaries and bonuses, but Daimler allowed its executives larger expense accounts. As stated above, it already seemed like the Germans were running the show, and the fact that they were six hours ahead of the Americans just magnified this feeling. Finally, key management on the American side resigned or announced their intent to leave. This aspect added to the stress of the changes and uncertainty brought on by the merger in the eyes of the American workers.

Did Daimler really practice comprehensive "due diligence" in terms of truly evaluating both what was really going on with Chrysler as well as the American car industry landscape? Did Daimler take a superficial view of how Chrysler was performing at the time, rather than consider, in-depth, the whole picture of Chrysler in terms of both its shaky performance history as well as its burdens? Did Daimler really take into consideration the legacy costs relating to healthcare and pensions for the Chrysler retirees? How about the fact that Chrysler possessed too many dealerships? All of these factors had burdened all the American car manufacturers, a situation that is very different than the case of auto manufacturers in other countries. It would appear to be logical that these issues should have been debated, and even quantified, rather than just being swept aside in the name of a strategic vision about consolidation and synergy.

Next, there were issues in terms of product offerings. For instance, there was overlap of product lines of the companies in terms of the two key SUVs, the Jeep Grand Cherokee and the Mercedes-Benz ML. Would such an overlap within the newly formed company, DaimlerChrysler, have a cannibalizing impact on the product lines? Furthermore, Chrysler did not turn out to provide the small car segment line-up that Daimler was seeking when making the merger deal. Specifically, the Chrysler brands were not in alignment with the Mercedes brand, and therefore, the small car offerings of Chrysler did not augment the luxury offerings of Mercedes to create a full suite of products. There were vast differences in the design and manufacturing processes, and thus it was difficult to achieve operational synergies. For example, Chrysler focused upon a front-wheel drive structure where as Mercedes used a rear-wheel drive structure. Furthermore, there was a concern for diluting the Mercedes brand, which created a dilemma between diluting the brand and saving money. For example, while money could be saved by sharing the vehicle platforms, should the Germans share their platforms without possibly diluting their brand? Mercedes engineers did not want to collaborate with Chrysler engineers, as Mercedes engineers felt that they were above those of the Chrysler brand. Senior management even feared technology transfer as they were concerned that too much talk regarding the Mercedes influence on Chrysler products would also dilute the Mercedes brand, as they had concluded happened with Ford and Jaguar where there was much talk about how the two lines shared many component parts.[21] Finally, there were differences in the way in which the Americans and the Germans marketed their vehicles. Chrysler's marketing strategy was designed for lower-priced, highly competitive markets, whereas Daimler's marketing strategy was designed for the high-end, luxury market.

The result of the merger was that neither of the two profitable companies benefitted. Not only were they not able to achieve their original goals that drove the merger initiative in the first place, but the merger actually created new problems. For example, while Daimler was busy helping Chrysler, BMW stepped in and seized the opportunity provided by the distracted competitor in the luxury marketplace to gain market share. Ironically, while Daimler was supposed to augment Chrysler in terms of quality, did Daimler's own quality deteriorate in the end? Another irony is that just a few months after the merger was complete, the German leader was already searching for yet another company to merge with since two of the key reasons for the first merger had not been met, the presence of an appropriate small car segment and access to the Asian market. This particular merger is a prime example where one plus one only equals, at best, one, but certainly not three, as originally hoped when making the deal. In the end, Daimler ended up having to pay another company, Cerberus, in order to remove the Chrysler part in 2007.

What are the takeaways from the failure of the DaimlerChrysler merger? One, companies need to learn from prior merger and acquisition failures. Daimler already had a track record of failed deals before it merged with Chrysler, from buying a stake in Cap Gemini Sogeti in 1991 to the purchase of Fokker in 1992. Both ventures were unsuccessful.[22] Clearly, Daimler never quite learned how to successfully plan and execute a merger or an acquisition, yet continued to pursue them. Second, if one gets into a bad deal, admit it and disengage. DaimlerChrysler continued on for years, despite the fact that it was clearly a failure. The third takeaway ties in with the first two, namely, to pay attention to the personality driving the merger. Daimler's CEO was known for his aggressiveness which is one of the reasons behind the string of prior mergers and acquisitions, as well as one of the reasons why DaimlerChrysler continued on—he would not admit his failure.[23]

On the other hand, the merger between Porsche and VW is likely to succeed due to a variety of reasons. First, both companies are German, which will eliminate the culture clash aspect that was prevalent in the DaimlerChrysler merger. Furthermore, not only are both companies of the same background in terms of nationality, but they also already have a history of working together, as far back as Ferdinand Porsche's involvement with the design of the VW Beetle in the 1930s or today with VW building the body and the diesel engine of the Cayenne. Finally, VW did have success bringing back the ailing Bentley, so it already has had experience dealing with other premium brands. The only concern is whether VW is wise enough to run Porsche at an arm's length. Porsche's turnaround resulted in a very direct and efficient corporate culture, very different from the far more bureaucratic organization and processes in VW.

Insights for Airlines

Airlines in the US have also had poor experiences with M&A initiatives. Just as in the case of Daimler and Chrysler, there were reasonably sounding rationales for the US Airways and America West merger. For example, America West had a strong route network in the Western region while US Airways' network was much stronger in the Eastern region. US Airways had a higher proportion of markets with higher yields while the America West market had much lower yields. America West had no presence in transatlantic markets, whereas US Airways did. However, despite these apparently valid arguments (and many others), the implementation has still not produced the expected benefits. The primary cause being the major difference in the two company cultures, particularly with respect to labor and the inability of management to integrate the two systems in a reasonable timeframe.

This point is comparable to the DaimlerChrysler failure. In both cases, the partners did not pay sufficient attention to the HR issues. For example, not only was the labor wage structure different in each airline, but the US Airways' workforce was much older, a fact that would cause serious problems in integrating the two seniority lists. There were also significant differences in the design and manufacturing processes between Daimler and Chrysler, making the achievement of operational efficiencies difficult. Likewise, there were enormous differences, for example, in the reservation and operational planning systems of the two airlines. In the final analysis in both cases, the HR and operational considerations were swept aside in the name of strategic vision about consolidation and synergy. On the other hand, the merger between Delta and Northwest will work because they confronted the HR issues right at the beginning.

What's Next for the Automakers?

Despite the problems faced by the US automakers, in particular, and the global automakers in the recent economic downturn, the future outlook is very positive for those willing to adapt. As Ronald Haddock and John Jullens point out in an article in *strategy+business*, the size of the global automobile market is enormous, given the desire for personal mobility in the "rapidly emerging economies."[24] Segments of the populations of these countries that can afford to buy vehicles is increasing rapidly. The opportunity lies in changing the business model, not only relating to manufacturing, but also to operations and marketing.

The American Automakers

In fairness, the automakers did begin to restructure their businesses through management reorganizations, some plant closings, appealing to the unions for changes in their agreements (such as a two-tier wage system) to regain competitiveness, and borrowing billions of dollars to finance major product innovations. In the case of GM, though, years and years of management arrogance, struggles with the unions, insular and disparate brands, as well as acceptance of the status quo could not all be undone in merely a few years time. It was just too little and far too late.[25] However, it is possible that Chrysler and GM could become stronger companies now that they have reorganized under bankruptcy protection. Now that GM and Chrysler have emerged from bankruptcy, Ford has one major competitive disadvantage. The debt obligations of the two bankrupt automakers have been drastically reduced. Ford, on the other hand, will need to pay its debt, making it less competitive. Similarly, American Airlines did not go through bankruptcy, and so also bears a greater financial burden than the competing, bankrupt airlines.[26] On the other hand, Ford appears to be in a better position having acknowledged earlier the failures in its marketing strategies. In the autumn of 2006, the new CEO brought in an experienced executive from Toyota. One of the new marketing executive's contributions was the realization that the dealerships needed to be involved for the marketing operation to succeed. This was a radical move given that it has been reported that Ford traditionally did not involve dealers in the creative process. Specifically, the new marketing executive stated that the dealers were going to be active in the company's focus groups, as who would know more about the customers than the dealers?[27]

Despite the ongoing, dismal situation in Detroit, authors Ron Harbour and August Joas state in their article in the *Oliver Wyman Journal* that they believe that Detroit could, in fact, make a comeback through the implementation of a few key elements, specifically innovation, passion, and speed of execution. In terms of innovation, the authors call for innovative technology that adds value. One example that they point out is Ford's success with its Sync voice-activated system in the Ford Focus. They also call for innovations that would augment a vehicle's functionality in these times of high energy costs. While they recognize that the Detroit Three were slow to get into the hybrid game, they still see potential for them to create other innovations, such as wireless access for carpooling passengers or more user-friendly bike carriers. In terms of passion, they call for the creation of small cars that instill just that. One example of a success story by foreign competition is the MINI, which has a phenomenal following. Finally, the authors state that the Detroit Three must increase their speed of execution in terms of their responses to shifts in market conditions. For example, while Ford is converting its SUV plants in response to the demand for small car production, these new cars will not be seen until 2010 or 2011, leaving a large time gap during which they will be missing out while the competition will be offering its products. In conclusion, the authors state that while the American automakers have talked a lot, it is now time to actually walk the talk and deliver.[28] The key is to restructure and reorganize, as Ford has already begun to do, and Chrysler and GM were forced to do through their bankruptcy filings, as well as remarket to be able to make money selling small cars by becoming the best value in that class.

As the authors of a recent article in *BusinessWeek* point out, Chrysler and GM could become stronger companies, but they still have many issues to face now that they have emerged from bankruptcy protection. First and foremost, they will have the baggage of their reputations. While the authors report that many Americans have long viewed the two companies' product offerings as dated and inferior, now the companies have the added burden of being viewed as drains on the US taxpayers; the companies have become referred to as "Government Motors." Furthermore, even though the companies have come out of bankruptcy as smaller, nimbler, and less bureaucratic entities that are able to compete in the global marketplace, they have emerged in an arena that is more hostile than any previous situation. Players in the auto game are keen to seize customers from Detroit. Consider the fact that the world's automakers plan to launch around 60 models in the US each year for the next five years.[29] Furthermore, both Kia and Volkswagen are planning on building new plants in the US, while Toyota has a factory in Mississippi that is currently slated for its Prius hybrid, but could also be used for the company's other offerings. Indian company Mahindra & Mahindra plans to begin selling its cars in the US in 2010 (for more on Mahindra & Mahindra, see Chapter 4).[30]

Can the US automakers compete in the US domestic market as well as in the changing global market? Several key attributes that need to be evaluated to determine the viability of the US automakers include the following.[31]

- Scale: to stay competitive in both cost and revenue.
- Quality: and product mix, present and pipeline.
- Flexibility: in execution, simplicity in platforms.
- Geographic Scope: to absorb risk.
- Partnering: skill or potential.

In reviewing the viability summary of Chrysler's plan, it is not surprising that it had no other choice but to go into bankruptcy.[32]

In reviewing the viability summary of GM's plan, the outlook seemed slightly better, but the company still ended up in bankruptcy.[33] What about Ford? Alex Taylor III provides some clues through his interview with Alan Mulally, Ford's CEO. Ford needs to accomplish the following:[34]

- Focus on the Ford brand.
- Compete in every market segment with carefully defined products.
- Market fewer nameplates.
- Become the best in quality, fuel efficiency, safety, and value.

The dire state of Detroit has had a downward spiral effect on the entire global economy. For example, while the Chinese consumers appear to have an affinity for Buicks, as discussed in the next section, it has also been reported that Chinese consumers have been concerned about the state of GM. Specifically, they were well aware that GM was on the verge of bankruptcy when it had to accept a bailout from the US government. An automotive blog in China raised the question, "Who wants to buy a high-ticket item from a potentially doomed company?"[35] Given that the company was forced to file for bankruptcy, it appears that their fears were substantiated. Furthermore, in Mexico, the slowdown in the automobile industry has forced some maquiladoras to lay off workers and/or cease operations. Similarly, the town of Bochum in Germany was fearing the closure of the local GM plant, where sedans and minivans are produced under GM's Opel badge.[36] Finally, the auto manufacturing corridor that encompasses Poland, Slovakia, and the Czech Republic (what has been referred to as "Detroit East") has too, been hit hard due to plunging vehicle sales worldwide.[37]

Back in the US, however, there is a new generation of entrepreneurs who are getting ready to design, manufacture, and market new kinds of autos in new ways. Take, for example, Henry Fisher, who is envisioning outsourcing literally every function (engineering, components, powertrain, and manufacturing), except design and marketing. These new entrepreneurs will not be burdened with employee constraining work rules or the high costs of post-retirement health care. Under such a business model (similar to what companies like Apple and Nike adopted), the new entrepreneurs could easily revitalize the auto industry within the US.[38]

The Changing Global Market

On a global basis, the market structure of the auto industries has been changing for some time, exemplified by the market capitalization of auto companies around the world. The two leading countries are now Japan and Germany. Out of a global total of US$633 billion in market capitalization (as of 4 June, 2009), GM's market capitalization amounted to US$376 million. According to the *BusinessWeek's* journalists, this value is less than Turkey's US$905 million Tofas Turk Otomobil Fabrikasi or Egypt's US$502 million Ghabbour Auto.[39]

As for the future, it is reported that the world is forecast to have three billion cars in 2050, compared to about 700 million today.[40] Consequently, while auto sales in mature markets (the US, Europe, and Japan) may seem to be saturated, sales in the emerging markets, especially the BRIC countries (Brazil, Russia, India, and China), represent a large potential for the auto industry. According to a special report in *The Economist*, there is enormous demand in the BRIC countries. Specifically, it has been reported that passenger vehicle sales in the BRIC countries are expected to surpass those in America. This statistic certainly illustrates the potential of these emerging countries as a source of growth for automakers who have basically tapped out the mature auto markets. While the economic crisis of the US is also felt in varying degrees in Europe and Asia, the BRIC countries are in a different position with respect to the credit crunch as the consumer behavior is different there. For example, levels of personal debt are relatively far lower, and a smaller proportion of vehicles are purchased via credit. The combination of three key aspects, rapid economic growth (relative to mature economies), favorable demographics, and social change, in these four countries are the main drivers of growth for the automakers. For example, GM's existence would be even more tenuous if it were not for the phenomenal growth in China. Specifically, the Chinese consumers appear to have an affinity for Buicks, as previously mentioned, which is interesting, given that the brand is flailing in the US. Twice as many Buicks were sold in China as were in the US last year. Moreover, in the first quarter of 2008, 65 percent of GM's sales were achieved outside of America, with a large portion of these sales attributed to the BRIC countries.[41] Furthermore, in some emerging markets (for example, in parts of China and Russia) there is even demand for high-end cars, despite high import duties.

It is important to note that each of the BRIC countries has distinct consumer preferences. For example, consumers in Brazil tend to prefer small cars, while the Russians appear to prefer SUVs. India values what leaders in the auto industry refer to as "frugal engineering," innovation and ultra-low cost. China, like Russia, has an affinity for larger vehicles. Furthermore, each of the BRIC countries has a different means of conducting business, as well as different attributes, and therefore foreign automakers must create individual strategies for each country in order to succeed. For example, until recently, the Chinese government required that foreign automakers set up joint ventures with Chinese partners if they wished to conduct business in China. Volkswagen was the first to form a joint venture, and did so in 1984 with the Shanghai Automotive Industry Corporation. General Motors followed in the 1990s. While foreign automakers no longer are required to operate through joint ventures, there are some positive reasons to do so, as pointed out by a Volkswagen executive. Specifically, it is a way to gain knowledge about culture and politics, rather than bear the risks of being an outsider. In India, companies such as Hyundai made the decision to conduct full manufacturing in that BRIC country, rather than incurring import duties of up to 90 percent. Other foreign competitors are following the same path, but Hyundai is one of the largest manufacturers. Russia possesses two striking attributes, having both abundant natural resources as well as drastically increasing disposable incomes (doubling in the past five years).[42] The Russian government implemented a law in 2005 in an effort to attract foreign manufacturers to assemble their products within the country. Specifically, foreign automakers have to build a factory in Russia meeting certain capacity and investment levels, but do not have to engage in joint ventures with local automakers, as they once did in China. Finally, the makeup of Brazil's auto industry is entirely different. Specifically, the auto market in Brazil is still run by the four automakers that have been there the longest—Fiat, Ford, GM, and VW. These four have operated in Brazil without local partners, and these four manufacturers held 80 percent of the market in 2007 with respect to light, passenger vehicles.[43]

Finally, there are also several joint lessons that can be learned from conducting business in the BRIC countries. First, companies must demonstrate a sense of commitment. For example, Fiat has excelled in Brazil because the company understood the market, trusted the local management, and invested heavily. Second, different countries and different companies require different business models; there is no one uniquely successful business model. For example, VW and GM have succeeded in managing joint ventures in China, but Japan has had to learn to play the game, as the Japanese have never been comfortable in such an arena. Third, companies must cater to the preferences and climates of each country when introducing vehicles from existing, mature markets. For example, both the Brazilian and Indian markets require small, fuel efficient vehicles that can endure poor roads, while Russians prefer SUVs that can tolerate the harsh winters. Even the basic powertrain options need to vary. For example, electric powertrains in China are desirable to reduce air pollution, ethanol-burning engines in Brazil (rich in sugar cane), and diesel engines in Russia (rich in oil).[44] Next, while the BRICs may be emerging marketplaces, magazines, the Internet, and increasing competition have greatly augmented buyers' expectations. Therefore, consumers in these countries will no longer accept inferior offerings. The final lesson is to not get carried away by the numbers. While there is huge growth potential in the BRIC countries, there is also much competition, which could prevent any one company from reaping enormous profits.[45] Not only will the Nano, introduced in 2009, compete with the older-generation Maruti Suzuki, but it will also compete with Ford's potentially game-changing Figo, a 2010 compact car designed specifically for the Indian market.

The Potential Game Changers

As authors Ron Harbour and August Joas were quoted earlier in the discussion as stating, they believe that a comeback could be made through the implementation of a few key elements—

specifically innovation, passion, and speed of execution. Some companies have begun to do just that to varying degrees from innovating products catering to the needs of particular countries all the way to changing the entire automotive business model. The following are a few examples of companies that are implementing Harbour and Joas' elements, and are well on the road to becoming potential game changers.

Fiat and Volkswagen

Fiat and VW are changing the auto game by offering vehicles with flex-fuel engines in Brazil. Fiat's offering, the Palio, and VW's Gol, are both equipped with a range of flex-fuel engines. These engines have the ability to run on any combination of gasoline and ethanol. The two automakers started this practice in 2003, and now almost every car produced in Brazil is fitted with such an engine. Ethanol has several positive attributes. First, it costs almost half as much as regular gasoline. Second, it helps shield the country from being a victim of volatile oil prices. Third, the ethanol industry provides jobs; it employs more than a million people in Brazil. Finally, ethanol is positive for the environment, as it produces lower emissions than gas and it requires little land (it only uses 2 percent of land currently in agricultural use). This investment in such engines by Fiat and VW illustrates the understanding of, and the desire to meet, the needs of the consumers of Brazil.[46]

Tata

Ratan Tata of the Tata Group noticed how whole families, including children, were traveling on motor scooters in India in the rain. Tata recognized the need to provide an offering that was (a) a safer means of transportation for these families, (b) would shelter these families from the elements, and (c) would be competitive with the pricing of a scooter. The process behind Tata's innovation will be discussed in detail in Chapter 4, *Innovating around Airline Realities*. Despite some issues, and therefore delays, with respect to the land for the factory, the Nano made its debut in March 2009. Reports indicated that the first 100,000 customers were chosen by lottery for the opportunity to purchase the base model Nano and an estimated 350,000 Indians entered the lottery.[47] The first delivery of the Nano vehicle took place in India in July 2009. There are also plans for a version with lower emission levels for the European market in 2011, as well as one with a relatively sturdier engine for the US in 2012.[48] It is interesting to put in some perspective the potential market for inexpensive autos, such as the Nano. According to one estimate (reported by the Society of Indian Automobile Manufacturers), 7.4 million two-wheelers and 1.5 million passenger vehicles were sold in India during 2008. According to the same source, Tata Motors has received orders for 200,000 Nano cars.[49] It is important to note that while Tata's decision to pursue the Nano concept appears to have been a success thus far, the decision to purchase British brands Land Rover and Jaguar from Ford have proven to have been less positive endeavors. Specifically, Tata Motors reported a quarterly loss as the result of sliding sales in its Jaguar and Land Rover brands. Moreover, Tata is seeking US$163 million in working capital to pump up the ailing British brands.[50]

Better Place

One company that could have a significant impact in terms of dramatically changing the automotive landscape is Better Place. Better Place, based in Palo Alto, California, and founded by Shai Agassi, is not a car manufacturer, but rather it offers a service (the service being "mobility miles") that can be applied to an electric car that a consumer purchases or leases from a regular name brand automaker. For example, Ghosn has already signed on for Nissan and Renault to offer electric cars. This is much like when a consumer purchases an iPhone from Apple, but then purchases the service for Apple's iPhone from AT&T. How does it work? Electricity is generated (utilizing renewable energy sources, including solar power and wind power, when possible), and then transmitted into a national car charging infrastructure. There is a service control center as well as stations located throughout the test area. Deals to test the model have been made with other countries including Australia, Denmark, and Israel, as well as the state of Hawaii and the San Francisco Bay area in the US.

Moreover, Better Place has entered into a partnership with a major taxi operator in Tokyo in which it is providing swappable batteries for a new fleet of electric taxis which will soon be in operation.

How is this different from traditional electric cars? Traditional electric cars can only travel a limited amount of miles, and then have to stop to recharge. Agassi's model has drivers stopping to switch out batteries, and then keep on traveling, a process that only takes a few minutes, similar to stopping at a gas station or a carwash. Moreover, the idea is to utilize robots to exchange the used batteries for fully charged batteries.[51] While the technology is just in test mode at this time, Agassi is on the cusp of totally changing the traditional automotive business model. At a time when American automakers have asked for handouts to pump into continuing their old business model and ultimately ended up going through bankruptcy anyway, Agassi is investing forward into a new business model. At the end of 2008, author Thomas Friedman put it this way: the taxpayers are being asked to provide bailouts for what he refers to as "Car 1.0," when we should be looking ahead at the race for "Car 2.0," which is exactly what Agassi and his company Better Place are doing. Finally, Friedman made the following analogy: Better Place's auto battery is like Apple's iPod, and Better Place's national plug-in network is like Apple's iTunes store. Both companies recognize that the physical device combined with the service offering represent a completely new business model in their respective industries.[52] While there are concerns, such as the cost of the infrastructure, (for example, battery-swapping stations cost US$500,000 each), Agassi believes that Better Place will succeed. He goes on to say that he has not seen any other viable plan for moving away from the world's dependency on oil.[53]

These new offerings totally change the automotive landscape. First, by keeping it simple, these automakers are able to reach many consumer segments, especially in the emerging markets, many who could not afford to own any car before these offerings became available. Tata's Nano and Renault-Dacia's Logan not only are affordable, but they are simple, so some consumers can even maintain the vehicles themselves, thus even lowering the cost to the consumer even more. Second, companies such as

Better Place totally open the door to personal mobility, allowing consumers to go wherever, whenever, without having to wait to recharge their vehicle; consumers can just switch out batteries and go where they need to go. Still to come are the "connected cars," that are connected for navigation and information, and to each other. The technology behind the "connected cars" could change the business model "from dynamic insurance and road pricing to car pooling and location-based advertising."[54]

Insights for Airlines

As Harbour and Joas point out, it may be possible for the US automakers to make a comeback through the implementation of a few key elements—specifically innovation, passion, and speed of execution. As pointed out earlier, there is a new breed of entrepreneurs getting ready to implement totally new business models to design, produce, and market new cars, the same is true for some US airlines. On a more global basis just as some traditional automakers could suffer a lot more because of game changer or format invaders such as Tata, Better Place, and Zipcar (discussed in Chapter 4), similarly some traditional airlines should also be concerned about game changers. It is true that some recent initiatives did fail, namely, Eos, MaxJet, Oasis Hong Kong and SilverJet. But there are many new initiatives started by entrepreneurs who have learned from the experience of those who failed. Some new airlines have already begun operations, for example, AirAsia X and V Australia, and some could enter the marketplace such as Ryanair in the transatlantic market.

The game-changing producers, whether automakers or airlines, will need to think of, and implement, a broader infrastructure. As with automakers, airlines would need to look beyond their own offering in terms of seats and networks, and consider the broader perspective of distribution as personal mobility becomes the key in attracting and retaining consumers. Finally, just as Haddock and Jullens point out that there are still good opportunities for the auto industry,[55] similar conclusions can be reached about the global airline industry, if the players were to adopt business models that pass the viability test with respect to scale, quality, flexibility, geographic scope, and partnering. Such leadership calls for willingness to change (as the US Government pointed out in the case of GM) to make the business truly consumer-focused, and to dramatically streamline management to facilitate the decision-making process on a much more timely basis (as discussed in the section on Business Agility in Chapter 5).

This is exactly what the new management at the new GM is proposing: to move quickly and decisively to focus on the following.

- Four core brands (with designs and technologies that matter to both consumers and the environment).
- A competitive cost structure, a cleaner balance sheet, and a stronger liquidity position.
- A culture focused on consumers and products.[56]

As with the new GM, the progressive airlines are recognizing, first, that "business as usual" is no longer an acceptable business model and, second, that they need to change their culture, a business activity that is under management control.

Notes

1 Osono, Emi, Shimizu, Norihiko, and Hirotaka Takeuchi, *Extreme Toyota: Radical Contradictions that Drive Success at the World's Best Manufacturer* (Hoboken, NJ: John Wiley, Inc., 2008).

2 Fear, Jeffrey, and Carin-Isabel Knoop, "Dr. Ing. h.c. F. Porsche AG (A): True to Brand?" *Harvard Business School* (Case Study), March 14, 2007, p. 2.

3 Edmondson, Gail, "Porsche's CEO Talks Shop," *BusinessWeek.com*, December 22, 2003.

4 Fear and Knoop, op. cit., p. 5.

5 Fear Jeffrey, and Carin-Isabel Knoop, "Dr. Ing. h.c. F. Porsche AG (A): True to Brand?" *Harvard Business School* (Case Study), March 14, 2007 and Fear, Jeffrey, and Carin-Isabel Knoop, "Dr. Ing. h.c. F. Porsche AG (B): Made in Germany," *Harvard Business School* (Case Study), April 3, 2006.

6 "What Makes for a Good Marriage? – Manufacturing – Renault-Nissan Alliance," *Automotive Industries*, November 2002.
7 "A Global Love Affair: A Special Report on Cars in Emerging Markets," *The Economist*, November 15, 2008, p. 8.
8 Ibid., pp. 14–5.
9 Maynard, Micheline, *The End of Detroit: How the Big Three Lost Their Grip on the American Car Market* (New York, NY: Doubleday, 2003), p. 11.
10 Ibid., p. 14.
11 Hoffman, Bryce G., "Jobs Bank Programs—12,000 Paid not to Work," *The Detroit News*, October 17, 2005.
12 Gaughan, Patrick A., *Mergers, Acquisitions, and Corporate Restructurings* (Hoboken, NJ: John Wiley, 2007), p. 156.
13 Rasmus, Jack, "Healthcare Special: VEBAs in the Auto Industry: How Companies Dump Union Negotiated Health Plans," *Z Magazine*, December 1, 2007.
14 Maynard, op. cit., p. 223.
15 National Highway Traffic Safety Administration website: http://www.nhtsa.dot.gov/portal/site/nhtsa/menuitem.43ac99aefa80569eea57529cdba046a0/
16 Kiley, David, "The Road to a Stronger Cafe Standard: U.S. Automakers Have to get Toiling to Meet New Fuel-efficiency legislation," *BusinessWeek.com*, updated 10:29 a.m. ET, Tuesday, March 25, 2008.
17 Solomon, Rob, Outrigger Enterprises Group, Presentation at the OSU International Airline Conference, Las Vegas, April 2009.
18 "Unstable Atom: General Motors Sells Saab," *The Economist*, June 20th, 2009, p. 70.
19 GM website, www.gm.com/news, "2009-07-10, The New General Motors Company Launches Today," July 10, 2009, p. 1.
20 Information was extracted from the following sources. Carroll, Paul B., and Chunka Mui, *Billion-Dollar Lessons: What You Can Learn from the Most Inexcusable Business Failures of the Last 25 Years* (New York, NY: Penguin Group, 2008); Gaughan, Patrick A., *Mergers, Acquisitions, and Corporate Restructurings* (Hoboken, NJ: John Wiley, 2007); Gaughan, Patrick A., *Mergers: What Can Go Wrong and How to Prevent it* (Hoboken, NJ: John Wiley, 2005); Lutz, Robert A., *Guts: 8 Laws of Business from One of the Most Innovative Business Leaders of Our Time* (Hoboken, NJ: John Wiley, 2003); Lynn, Barry C., *End of the Line: The Rise and Coming Fall of the Global Corporation* (New York, NY: Doubleday, 2005); Rosenbloom, Arthur H. (editor), *Due Diligence for Global Deal Making* (New

York, NY: Bloomberg Press, 2002); Stahl, Günter K., and Mark E. Mendenhall, *Mergers and Acquisitions: Managing Culture and Human Resources* (Stanford, CA: Stanford University Press, 2005); Vlasic, Bill, and Bradley A. Stertz, *Taken for a Ride: How Daimler-Benz Drove Off with Chrysler* (New York, NY: HarperCollins, 2001); Waller, David, *Wheels on Fire: The Amazing Inside Story of the DaimlerChrysler Merger* (London: Hodder & Stoughton, 2001).

21 Carroll, Paul B., and Chunka Mui, *Billion-Dollar Lessons: What You Can Learn from the Most Inexcusable Business Failures of the Last 25 Years* (New York, NY: Penguin Group, 2008), p. 180.

22 Waller, David, *Wheels on Fire: The Amazing Inside Story of the DaimlerChrysler Merger* (London: Hodder & Stoughton, 2001), p.37.

23 Gaughan, Patrick A., Mergers: *What Can Go Wrong and How to Prevent it* (Hoboken, NJ: John Wiley, 2005), pp. 315–6.

24 Haddock, Ronald, and John Jullens, "The Best Years of the Auto Industry are Still to Come," *Strategy+Business*, Issue Number 55, Summer 2009, pp. 36–47.

25 Holstein, William J., *Why GM Matters* (NY: Walker Publishing, 2009).

26 Ingrassia, Paul, "How Ford Restructured Without Federal Help," *The Wall Street Journal*, May 11, 2009, p. A19.

27 "Detroit's Race Against Time," *The Economist*, August 9th–15th, 2008, pp. 60–61; Kiley, David, "The Fight for Ford's Future," *BusinessWeek*, August 11th, With a Huge Overhaul of its Factories, the Price of Fuel May Come Crashing Back 2008, pp. 40–43; Muller, Joann, "Whipsaw: Just When Ford Catches up to $4 Gas Down. God Forbid," *Forbes*, September 15th, 2008, pp. 38–40.

28 Harbour, Ron, and August Joas, "How Detroit Can Stage its Next Comeback," *Oliver Wyman Journal*, Fall 2008, pp. 26–29.

29 According to J.D. Power & Associates, as reported in *BusinessWeek*: Welch, David, and David Kiley, "The Hard Road Ahead for Government Motors," *BusinessWeek*, June 8, 2009, p. 31.

30 Welch, David, and David Kiley, "The Hard Road Ahead for Government Motors," *BusinessWeek*, June 8, 2009, pp. 28–32.

31 Solomon, op. cit.

32 http://online.wsj.com/public/resources/documents/Chrysler-Viability-Assessment20090330.pdf

33 http://online.wsj.com/public/resources/documents/GM-Viability-Assessment-20090330.pdf

34 Taylor III, Alex, "Fixing Up Ford," *Fortune*, May 25, 2009, pp. 44–51.

35 Welch, David, "GM Hits a Wall in China, Too," *BusinessWeek*, February 16, 2009, p. 32.
36 Ewing, Jack, "What's Dragging Europe Down," *BusinessWeek*, March 9, 2009, p. 36.
37 Ewing, Jack, "The Car Slump Slams 'Detroit East'," *BusinessWeek*, June 29, 2009, pp. 52–4.
38 Muller, Joann, "Automakers' Gold Rush," *Forbes*, June 8, 2009, pp. 70–77.
39 "The Outlines of the New Global Auto Industry," *BusinessWeek*, June 22, 2009, p. 17.
40 "Detroitosaurus wrecks," *The Economist*, June 6, 2009, p. 9.
41 "A Global Love Affair: A Special Report on Cars in Emerging Markets," *The Economist*, November 15, 2008, p. 4.
42 *The Economist*, November 15, 2008, op. cit., p. 7.
43 *The Economist*, November 15, 2008, op. cit., p. 6.
44 Haddock and Jullens, op. cit., p. 39.
45 *The Economist*, November 15, 2008, op. cit., pp. 3–9.
46 *The Economist*, November 15, 2008, op. cit., p. 7.
47 Srivastava, Mehul, "What the Nano Means to India," *BusinessWeek*, May 11, 2009, pp. 60–1.
48 "Hello, Nano," *BusinessWeek*, April 6, 2009, p. 8.
49 Karmali, Naazneen, "Every Village, Every Home," *Forbes*, June 8, 2009, pp. 80–81.
50 Fontanella-Khan, James, and John Reed, "Tata seeks £100 m boost for UK brands," *Financial Times*, September 1, 2009, p. 16.
51 Hamm, Steve, "The Electric Car Acid Test: Shai Agassi's Audacious Effort to End the Era of Gas-powered Autos," *BusinessWeek.com*, January 24, 2008 and "The Electric-fuel-trade Acid Test," *The Economist*, September 5th, 2009.
52 Friedman, Thomas L., "While Detroit Slept," *The New York Times (nytimes.com)*, December 10, 2008.
53 "Electric Evangelist: Can Shai Agassi of Better Place, an Electric-car Company, Honour his Grand Promises?" *The Economist*, May 2, 2009, p. 68.
54 "The Connected Car," in a Special Section, 'Technology Quarterly' *The Economist*, June 6, 2009, p. 18.
55 Haddock and Jullens, op. cit., pp. 36–47.
56 GM website, www.gm.com/news, "2009-07-10, The New General Motors Company Launches Today," July 10, 2009, p. 1.

Chapter 3
Learning from Other Successes: The Customer Experience Industry

Nearly every company seems to believe that its customers are placed as a top priority in its organization. Specifically, it has been reported that more than 80 percent of companies believe that they deliver a superior customer experience.[1] However, in reality, studies show that only 8 percent of their customers agree,[2] and the situation is likely to become worse as companies continue to feel the pressure to reduce their costs. Yet, customer experience has been, and continues to be, a great way for a business to distinguish itself in the crowded marketplace.

Why do companies who claim to be customer-focused only pay lip service? These questions are particularly relevant within the airline industry, and especially within the US airline industry where, with few exceptions, much improvement is needed (even though some elements of the passenger experience are not within the control of airlines). While there are a number of problems experienced by passengers due to factors outside the control of airlines (for example, the limitations of the aviation infrastructure and facilities relating to airports, the ATC, and immigration facilities, as well as lengthy delays and processes related in government-implemented security procedures), there are still numerous areas in which airlines can make improvements.

This chapter begins with some insight into the first main driver behind consumer experience, the customer groups (for example, the rise of new potential consumer groups and consumer behavior in times of economic downturns) and ends the first part with some insights for airlines. The chapter then provides an overview of the next two main drivers of customer experience, employee capabilities and corporate processes, systems, and incentives. The chapter ends with three case studies of companies (non-airline businesses) who have excelled in their consumer experience arena, followed by some takeaways for airlines.

As illustrated in Figure 3.1, the drivers behind customer experience can be separated into three main components: the customer groups, employee capabilities, and processes, systems, and incentives.[3] While the topic of customer experience is certainly not new, it is becoming imperative to review and re-evaluate it due to several factors. First and foremost, despite all of the discussion surrounding the improvement of the customer experience over the past decade, many consumers still feel that they are receiving poor service in many industries, especially the airline industry. Additionally, the Web has created incredibly informed consumers, who are able to research every aspect of an airline's products or services (including an airline's competitors), and obtain ratings and feedback from fellow consumers on all of the above. Search engines and social networking sites have totally changed the consumer and organizations must adapt.[4] Some industries, such as the telecommunications industry, have taken social technology a step further, specifically in recognizing that social network analysis is becoming an important tool for predicting and influencing consumer behavior.[5] For more on Business Analytics, see Chapter 5.

Customer Groups

One experience does not fit all. As Bruce Temkin of Forrester Research recognizes, the exact same experience can be satisfactory for one person, but totally unsatisfactory for another, and this is true even within one region, one culture, or one generation. Therefore, experiences need to be customized to individuals, or at least designed for customer segment groups as opposed to being directed at all customers in general. Today, technology enables a company to understand its customers on a personal level. The information provided in this chapter goes into more detail than the basics on consumer demographics covered in the previous book.[6] These details might be helpful for airlines not only from the viewpoint of developing new frameworks for customer

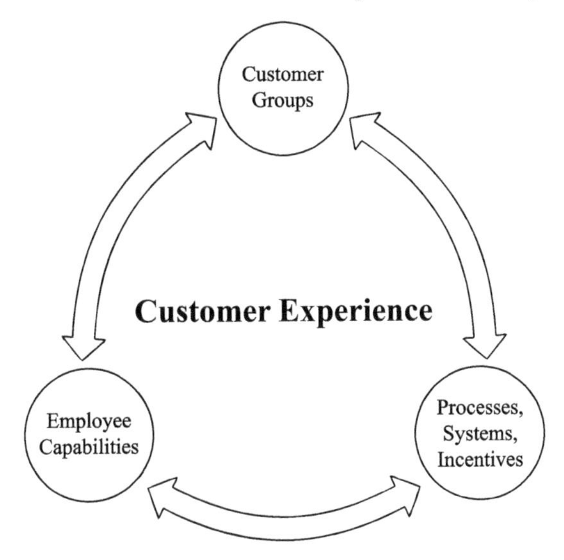

Figure 3.1 The Three Drivers of Customer Experience

experience, but also from the viewpoint of future employees. The following is a discussion of the different consumer groups in North America today, with an emphasis on the Millennials, in particular.

Millennials

Size and Geographic Concentration One new member of the consumer portfolio that is currently emerging in the marketplace is the Millennials. While various analysts define the parameters of the Millennial generation slightly differently,[7] they all agree that the Millennials are the largest generation, much larger than the Baby Boomer generation. This emerging generation is predicted to have siginificant spending power and influence in the world, whether they are still in school or are entering the workforce. Research indicates, for example, that the total spending power of Millennials is at least US$200 billion per year.[8] Some general attributes used to describe the up-and-coming Millennial generation follow (Millennials with respect to consumer experience will be addressed later in this discussion); however, generations cannot be totally and clearly separated with beginning and endpoints, and certainly not all individuals that fall within this age bracket portray all of these characteristics. This discussion is meant to provide readers with some insights about a new generation and its possible implications on businesses, especially airlines.

It is important to note that the Millennial phenomenon is most prevalent in the US, and, to an extent, Canada, but not so much in the rest of the world. In terms of the social technology aspect of the Millennial generation, it has been observed that German youth actually do not spend as much time, for example, on Facebook, due to privacy concerns and also because Facebook has become too complex. However, recent articles indicate that Facebook is becoming relatively more popular in Europe, especially in Italy where Facebook grew by more than 2,700 percent (all users) over the past year (2008–9).[9] In any case, since the significant growth is predicted to be in countries such as China and India, it is important to consider the attributes, consumption behavior, expectations, etcetera of the same age groups in those countries.

Aptitude for Technology One of the most influential characteristics of the Millennials is their aptitude for technology. They have grown up with technology surrounding them, and, therefore, are extremely savvy with respect to its use. They are not only heavy users of technology, but they expect all those with whom they interact to be equally as able. They are avid learners, but prefer to learn using technology rather than through traditional means. For example, as one college official points out, they are willing to navigate through 1000 pages of online content, but complain about reading a 20 page case study in paper form.[10]

Millennials are well connected, mostly due to social networking websites such as Facebook, LinkedIn, and MySpace. The Internet has made the whole world a part of the Millennials' community; the Millennials are able to create connections with people who have similar life experiences and values even if they live on the other side of the world. They also enjoy sharing everything via the Internet, for example, pictures on Facebook, videos on YouTube, and their opinions on blogs. They fully embrace mobile media, especially cell phones and iPods, and they are not just buying these products and services, but are actually emotionally bonding with them. Text messaging, instant messaging, and downloading pictures, music, and ringtones are all instantaneous means of communication with almost anyone, anywhere in the world. In addition, many video games involve players competing against each other online, and these players can be located almost anywhere in the world. According to one article, Millennials "not only value connectedness, they are defined by it."[11]

Given that they are so technologically oriented, it is not surprising that Millennials are extremely proficient at multitasking. They can be seen walking down the street listening to their iPods while texting away on their cell phones, while also enjoying a cup of Starbucks. If sitting at a desk, add in being on Facebook, MySpace, and YouTube to the list. Ron Alsop, author of *The Trophy Kids Grow Up*, ranks technological savvy and the ability to multitask as two of the greatest attributes of the Millennial generation. Alsop recognizes that this group desires to be connected continuously around the clock, while at the same time they wish to be engaged in numerous experiences. The combination of these two attributes makes this group very demanding.[12]

Lifestyles: Expectations and Attitudes Millennials are strongly committed to the idea of innovation, and are creative, flexible, and full of energy. They love variety, and possess an entrepreneurial spirit. They do not work well within a rigid structure or set of rules, which is not surprising considering their flexible and entrepreneurial attributes. Their top priorities combine career and community service, and they are considered a generation of action. When Hurricane Katrina devastated the state of Louisiana in the US, Millennials flocked there to provide help. In addition to being community service oriented, they are open minded, supporting racial blending, immigration, and the general concept of "live and let live." They are deeply concerned with equality and the common good, which ties in with their strong sense of community service and volunteerism. This generation is extremely pro-environment.

It is important to note that while this generation is career oriented and is keen on getting into the right college and company, as a whole they do place their lifestyle before the job. For example, a Millennial will chose a location where she would like to live after graduation, and then start a job search in that location. This is extremely different from older generations who believed in going "wherever the job may be." However, as recent articles in *The Economist* address, the Millennials will have to adjust their expectations in the near term given the current economic downturn. Specifically, there are far fewer jobs available for Millennials to be moving to and from. On the other hand, the articles also recognize the strengths of Millennial employees in such economic times. Specifically, Millennials are flexible in the sense that they are willing to take on a new role (within the company), travel for work, or even move to a different location if necessary. Their mobility and lack of commitments outside of work are key attributes in such times. Furthermore, their technological aptitude and drive for innovation can potentially create savings for their employers. For example, it was reported that one Millennial employee at Best Buy, a US retail electronics company, put together a mobile phone version of the corporation's website "just for fun" in his free time. Finally, the articles also point out that other desires of the Millennials in terms of the workplace, such as increased feedback and coaching, as well as implementing and utilizing more technological communication tools, are worthwhile enhancements regardless of the state of the economy.[13]

Consider, for a moment, just one characteristic of the Millennials, namely, their high expectations. Many analysts attribute this level of expectation, or sense of entitlement, to the fact that this generation was never denied much as children. They were coddled by their parents as well as their teachers. They were made to feel special and told that they could do anything, and now they expect those results. Their special connection to their constantly intervening parents continues way beyond childhood and even their high school years. These "helicopter parents" (named as such due to their tendency to hover over their Millennial offspring) are thoroughly involved in getting their Millennials into college, through college, and into the workplace, much to the chagrin of many university admissions staff and employers.

In the workplace, Millennials expect, or perhaps even demand, work-life balance and a flexible schedule. They desire a casual atmosphere where they can be comfortable and can be themselves. They want the ability to stay connected even while at work, listening to their iPods and interacting with their friends and acquaintances on various social networking sites. They expect to use the same technology at work that they have been using at school (or during most of their lives, for that matter). For example, they want to be able to email or text their bosses rather than having to call them. They want to be able to use technology to conduct meetings rather than have to have the traditional face-to-face encounters. They do not view technology as a luxury, but rather as a necessity in conducting business. Thus, they expect to have a laptop and a cell phone to have the ability to work after traditional hours and in other locations. This generation knows that they have to work more hours than the traditional 9–5 schedule typical of their parents' generation, but wish to decide where and when they will put in the work time. Ubiquitous computing ability yields productivity through working at night and on weekends, and, therefore, they see these devices as necessary tools, not perks as they are thought of by older generations, who often view this group as too demanding or even disrespectful of seniority.

Millennials dislike hierarchy, and, therefore, not only want to be able to be promoted quickly, but also to immediately be treated as colleagues rather than junior or entry-level employees. They do not consider it unreasonable to expect to have the opportunity and access to share their ideas with senior executives all the way up to the CEO. They desire constant feedback, way beyond the traditional annual review or semi-annual meeting. Most of all, they want to have meaningful work. They want to make a difference. At the same time, however, they also expect to have fun in the workplace. Millennials desire a true meritocracy that rewards those who truly perform rather than those who just put in their time.

This is a definite point of distinction between this generation and the older generations who believe that if you do your time, you will be promoted. The older generations are taken aback by what is perceived as a lack of loyalty by the Millennial generation.

Millennials tend to change jobs often in their search for the ideal career. They want to learn as much as they can, and have as many different experiences as possible. Many are delaying the milestones of adulthood—marriage, children, and home ownership—to do so. This generation always has "one foot out the door;" if a job does not prove to be fulfilling, they will simply leave. Many of this generation may even end up leaving the corporate world entirely, and become entrepreneurs instead. Their company disloyalty partially stems from the fact that Millennials have had family and friends laid off due to corporate downsizings or outsourcing of jobs to other countries. Millennials know that job security is a luxury of the past, so they want to be as prepared as possible for potential career changes. From a Millennial's point-of-view, their lack of loyalty is really a strategy of survival. Their ability to obtain as many transferable marketable skills as possible is the key to their survival in the current uncertain workplace.

As with their work, as mentioned earlier, Millennials value things that have meaning. It is all about the experience. Authenticity is also a big part of marketing brands to Millennials, and it is about "being open to the feedback from the market and being honest in the communication."[14] This generation will not accept products and services that are not authentic. Responsibility is also a key issue with the Millennials; they are greatly concerned with the manner in which a company conducts business and in how the company interacts with, and contributes to, the communities that it serves. Furthermore, since one of the greatest characteristics of the Millennial generation is their technological aptitude, businesses are learning how to use interactive tools if they want to market to Millennials. Again, Millennials expect all those with whom they interact to be equally as savvy, utilizing the same electronic mediums as they do.

Finally, long-term anything is out for this group. Short-term is their reality, as they have a compressed view of time, and technology makes them very impatient, as they are used to getting everything right now. It has been said that the Millennials may be the highest maintenance workforce in history, as they desire everything right now—coaching, feedback, promotions, recognition, and training. However, it has also been said that if management can find a way to tap into their energy and to mentor these up-and-coming employees, this group has the potential to become extremely high producers. To do this, management will need to implement an approach that is more engaged, timely, and focused on the short-term in addition to the long-term.

Adaptations by Businesses How have businesses, including airlines, adapted to the Millennials? Alsop has identified several companies that have already jumped on the bandwagon in terms of targeting the Millennials. Specifically, some companies are attempting to connect to the Millennials' needs in the workplace, such as work-life balance and flexibility. For example, Google has catered to the need to merge work and personal life by providing everything from car washes and oil changes to hair salons and massages to dry cleaners and bike repair services at its headquarters. Oliver Wyman, a management consulting firm, has introduced a program to cater to the Millennials' desire to travel and pursue their interests. Specifically, employees may choose to work 10 or 11 months of the year at a reduced salary, and may take sabbaticals of up to six months. The company will retain the employee while she pursues her interests, which could include anything from producing a film for the Montreal World Film Festival to playing for a professional football club.[15]

Other companies such as Merrill Lynch and Ogilvy have made attempts to connect with the Millennials on a different level. Specifically, certain divisions of these companies have tried to cater to the "helicopter parents" of the Millennials by creating "Bring Your Parent to Work Day" in an effort to help the Millennials' parents feel involved, understand what kind of work their "children" are doing and learn more about the company. These parents are jumping at the opportunity to be involved, and in the case of Merrill Lynch, some parents have traveled from as far as Nigeria to participate in the company's program. Other companies, such as Office Depot, are launching websites with special parents' sections in an attempt to address parents with recent graduates entering the workforce. Finally, Alsop indicates that Enterprise Rent-A-Car might in fact be one of the most accommodating employers as they are happy to interact with parents in terms of information regarding the company as well as job opportunities.[16]

Since the Millennials tend to place the greatest emphasis on their overall lifestyle, businesses are realizing that they expect more than just good products and services. With respect to travel, however, it seems that only the opposite is currently and typically available to all generations. The actual experience of traveling is often awful, at least in the US, but passengers must tolerate the poor experiences in order to achieve another objective, such as attending a business meeting or starting a vacation. Millennials expect great experiences from everything they purchase, whether it is an item from a store or entertainment in a theater. The bottom line is that Millennials will not tolerate something that is not exciting or interesting. According to one business writer, the big trend will be a shift from experiences that are just fun, to experiences that are also life changing in the sense of altering the way in which people view the world. Sheahan gives the example of Cirque du Soleil. Cirque du Soleil has not just created a great entertainment experience, but rather they have created a life changing experience in the form of an artistic and athletic performance.[17] For additional information on Cirque du Soleil, see the case study in Chapter 4.

How do businesses get the attention of the influential Millennials? To answer the question, according to one article (in ResourceInteractive's *Litmus*), there are five key rules to keep in mind when marketing to Millennials.

- First, as mentioned above, be authentic. This generation not only wants to do business with authentic companies providing products, but they also desire constant new product offerings. For example, Millennials love eBay, because it is "always fresh."

- Second, Millennials want to be heard. The Millennials want to interact with companies that listen to them and involve them in their decisions. For example, this group likes Amazon.com, because it offers user reviews, and therefore is perceived as a company that listens to its customers.

- Third, Millennials thrive on creativity, as mentioned in the previous section. They enjoy limited editions, product exclusives, and so forth. Again, they like excitement, freshness, and originality, so frequent updates to offerings greatly appeal to this group.

- Fourth, Millennials want control over the purchase interaction. As stated in the previous section, they are impatient. They want everything now. Apple's iTunes Store is an example of an offering that caters to this need of everything to be on the purchaser's terms. Specifically, an individual can select her songs, download them, and enjoy listening to them through her iPod at any time of the day or night. Not only is it better than a brick-and-mortar store that closes, but even when one purchases CDs via the Internet, one has to wait for the purchase to be shipped. Apple's iTunes store is truly 24 hours; one can not only purchase, but also enjoy the purchase instantly.

- Fifth, Millennials want to be entertained. Viral events that drive buzz and can be shared are favored by this generation.[18]

IDEO, a design consultant firm based in Palo Alto, California, was hired by PNC Financial Services in 2007 to study the Millennial generation and formulate an initiative directed at targeting this new group of banking customers. The analysis produced two major findings: (1) individuals in this age group tend to find banking websites to be "clunky," and (2) in general, they need assistance managing their finances. Given these findings, the "Virtual Wallet" online program was implemented. Features of the online program include the ability to "drag" money from one account to another all within one screen, and to view balances on a calendar basis instead of a ledger to plan for bills and analyze spending habits. One simple, but key, feature that one Millennial reported to have attracted her to the program is the ability to obtain balances via text message. Furthermore, the program is designed to grow with the Millennial generation. For example, as the Millennials mature and have more advanced banking needs, the Virtual Wallet program will grow, too, in terms of offering loan and investment tools.[19]

More recently, Ford Motor Company has incorporated the Millennials into the introduction of the company's new generation of global, small car offerings. Specifically, Ford implemented a clever program in which 100 young consumers were loaned the new Ford Fiesta for six months in order to test out the new product and generate buzz in anticipation of the launch of the vehicle in the North American market in 2010. The program is a part of the "Fiesta Movement," in which 100 Millennials were chosen out of over 3,000 applicants to drive the European Ford Fiesta, and share their experiences with the product via social networking sites, including Facebook and YouTube, as well as with Ford. Ford executives recognize that the Millennials are the up-and-coming generation of automotive consumers with an estimated 70 million drivers in 2010. What better way to introduce them to, and connect them with, the brand than by capitalizing on some of their key attributes such as sharing experiences via social media?[20]

Other Generations

Generation X While the previous section focused on the Millennial generation, it is important to note that Millennials may be influenced, to some extent, by their older siblings who are members of the preceding generation, Generation X. Specifically, Gen Xers were generally born between the mid-1960s and the mid-to-late 1970s.[21] They represent the smallest segment of the US population (the Millennial population is three times the size of the Generation X population).[22] To this point, some members of this generation feel that they are forgotten, especially by the media, who tend to focus on either the Baby Boomers or the Millennials, thus skipping over the Gen Xers. It has been reported that Gen Xers tend to be independent and possess the ability to fend for themselves. This could be partially due to their generation having the highest number of divorced parents or both parents in the workforce, compared to prior generations.[23]

Baby Boomers It is also important to consider the Baby Boomers (born between the mid-1940s and the mid-1960s) as they, too, are still certainly active consumers, and represent a significant group in the marketplace (one report as of mid-2008 indicates that there are almost 100 million consumers in the US who are age 50-plus[24]). However, many reports indicate that Baby Boomers have been forgotten when it comes to advertising. Specifically, advertisers continue to prize youth; only roughly 10 percent of advertising is directed specifically at the 50-plus market.[25] This seems counterintuitive given that Baby Boomers are reported to be, as a group, the most affluent Americans. While their investments have certainly taken a hit due to the economic downturn, this group still has money to spend. Consider, for example, that more than half of the cars sold in the US each year are purchased by consumers in the 50-plus age bracket, yet only 10 percent of advertising dollars are directed at promoting cars to this group.[26]

However, a select few companies have recognized this, and are capitalizing upon it. For example, Anheuser-Busch has advertised its low-cal, low-carb Michelob Ultra beer in the AARP magazine, featuring fit, active 50-plus individuals participating in activities such as swimming, kayaking, and biking.[27]

Consequently, it is important to address each generation's unique attributes and needs. Gravett and Throckmorton, authors of *Bridging the Generation Gap*, stress the need to tailor recruiting methods in the workplace to each generation, but this concept can also be extended and applied to marketing to, and attracting, consumers. What are the distinct needs of each generation? How do they like to be contacted? How do they interact with businesses? As examples, Gravett and Throckmorton point out that employers who only offer benefits such as daycare, maternity leave, and tuition reimbursement are missing an opportunity to attract potential older employees who possess great experience. They stress that the "hook" that attracts individuals to your business will have a different shape for each of the populations.[28] Are airlines missing out on the needs of a particular generation of consumers, not to mention potential employees?[29] There are difficult decisions to be made. For example, many airlines are currently evaluating the provision of Wi-Fi services in aircraft. Obviously, most of the Millennials and a significant segment of the older generations are likely to have the need to be connected online on board the aircraft. As for airlines, making Wi-Fi available would certainly generate significant ancillary revenue. On the other hand, there is also a segment of the older generation that is likely to consider this service not only a menace and a distraction, but also a potential problem with respect to security relating to the information being viewed by passengers.

Consumer Behavior in Economic Downturns

Obviously, businesses recognize that economic downturns can be an opportunity to truly analyze consumer behavior and realign a business' strategy accordingly. Some business writers call for a complete analysis through what they refer to as "ground-level intelligence." This involves getting at the heart of consumer behavior, and recognizing patterns and trends. For example, Wal-Mart ascertained a pattern in consumer behavior in which, for the first time ever, sales of baby formula coincided with twice-monthly pay periods, indicating that consumers were truly strained, and thus altering their purchasing behavior of certain goods as a result.[30]

Quelch and Jocz, authors of an article in the *Harvard Business Review* also recognize that during economic downturns, companies need to understand the shift in consumption behavior by consumers. While Quelch and Jocz recognize that no two downturns are exactly the same, there are patterns that have been identified in terms of new customer segments that emerge during such times. They note that while companies tend to segment consumers demographically, this method can be less relevant than segmenting consumers psychologically during times of economic downturns. Specifically, they suggest segmenting consumers into four groups, ranging from those who are hit the hardest and, therefore, are drastically reducing all spending, to those that live for the here and now, and are continuing on as usual. The first three groups tend to reevaluate their purchase choices during an economic downturn, while the last group's patterns remain primarily the same. Members of this last group tend to be younger, renters, less concerned with savings, and spend more on experiences rather than physical items, except when it comes to electronic goods. It is imperative during such times to continue to invest in business intelligence, following how consumers reevaluate priorities, spending, value, and brands. (For more information on Business Intelligence, Pricing, and Value, see Chapter 5.) Quelch and Jocz stress that those companies that consider and embrace consumer behavior psychology, and adjust their marketing accordingly, will be most likely to be successful during economic downturns as well as in the times following. Dell is an example of a company that appears to have attempted to customize its messages to appeal to each of the above consumer segments and their specific needs.[31]

An article in *The Economist* also notes the impact that such a severe economic downturn will have in terms of consumer psychology going forward. Specifically, the article identifies three main areas that companies will need to be concerned with as they conduct business with consumers. First, companies will need to be more transparent than ever with their consumers, as the failings of many businesses, as well as the severity of those failings, has consumers at an all time high level of suspicion and concern. Second, companies need to be empathetic with their consumers. An article in *BusinessWeek* illustrates this concept through several companies who have already implemented practices to address this need. For example, Hyundai has implemented programs which include taking a car back during the first year of ownership with no harm to the customer's credit if the customer loses their job (for more on Hyundai's program, see Chapter 5). Ford and General Motors have also implemented similar programs. Another example is jetBlue, who launched a program in which the airline would refund the cost of tickets purchased during specified dates if a passenger lost her job. FedEx Office (formerly FedEx Kinko's) offered one day where those searching for a job could have their resumes printed for free.[32] The final point ties in with the first one. This economic downturn will most likely increase the use of social technology by consumers as they will look to fellow consumers for insight on products and services rather than trusting companies and brands themselves.[33]

In terms of corporations, consumer electronics retailer, Best Buy has been highlighted as a "creative retailer" in the current economic downtown. Specifically, Best Buy has recognized a shift in consumption patterns, and is altering its offerings accordingly. The company's analysis determined that in such times, consumers are spending relatively less on new gadgets, but have a need to make the ones that they already have last longer. Best Buy is capitalizing on this shift in consumer needs by expanding its technology services division, including its Geek Squad tech support team. Best Buy recognizes that by fulfilling this need of serving a customer over the life of their purchase rather than just for the initial transaction, the company can distinguish itself in the marketplace.[34]

A second article in *The Economist* also addresses the impact that such a severe economic downturn will have in terms of future consumer psychology. Specifically, unlike the book by Silverstein and Fiske of a few years ago, *Trading Up: The New American Luxury*, in which Americans, even those who were not wealthy, were trading up for luxurious products and services, Americans are now trading down. They are doing without even the smallest luxuries, as they are in fear of losing their jobs, if they have not already done so, and their home values, along with their investments, have plummeted. The emphasis is now on thrift, and, therefore, it is not surprising that the only two stocks that rose in 2008 in terms of the Dow Jones Industrial Average were McDonald's and Wal-Mart. One clear example of people trading down is the reduction in sales at Starbucks. Some people are trading down the US$4 latte at Starbucks to a lower-priced coffee at McDonald's or Dunkin' Donuts.

An article by Flatters and Willmott in the *Harvard Business Review* substantiates this emphasis on thrift, noting that even affluent consumers are following this trend, although they may not need to do so. The authors recognize that many consumers are becoming disenchanted with excessive consumption, and are moving toward a way of life that is more wholesome and less wasteful. They note that this trend of thrift has not only become acceptable, but potentially even fashionable given the recent, severe, economic climate. This trend of discretionary thrift, as Flatters and Willmott call it, ties in with another trend that they address, namely, a demand for simplicity. This trend was becoming prominent even before the recession, as consumers were feeling overwhelmed and inundated by not only the abundance of choices, but by always being connected. Companies that have capitalized on the simplicity trend include *Real Simple* magazine, aimed at making everyday life easier with recipes, home organization tips, etcetera, and Apple, with its sleek, minimalist, easy-to-use iPod.[35]

Insights for Airlines

Since it has been reported that Millennials have a significant influence on travel plans within their households, it is important to consider the emerging trends in the travel sector. Ypartnership lists the following trends.

- Value is top priority. In light of the recent economic events, consumers are expecting even more for their dollar. This does not mean that they will necessarily choose the least expensive option, but rather that the consumer will search out the best deal. All-inclusive travel services will become more popular as consumers want to have more control over what to expect in terms of the total cost of their travel.

- In line with cost avoidance measures, both business and pleasure trips will be shortened in length, thus travel providers will need to make up the resulting gap in revenue by increasing the volume of consumers.

- Consumers will change the manner in which they use the Internet. Specifically, consumers crave sites that do the price comparison shopping for them. Farecast and Kayak are two such sites, and fly.com is an example of a site that enables comparison of value, not just price.

- More travelers will use their mobile phones for travel planning and purchasing, given the increasing usage of such devices.

- Consumers are increasingly using vacations as a means to celebrate milestones, such as anniversaries, birthdays, graduations, and retirements. Disney has already capitalized on this trend by implementing the promotion of free admission to any Disney park on one's birthday in 2009.

- Consumers prefer to utilize travel service providers who are "green," although they, for the most part, do not wish to pay more to utilize green suppliers.

- The makeup of the American consumer is drastically changing; diversity will become a greater factor in the marketplace.[36]

Consequently, the Millennials are likely to be characterized more by technological aptitude and interests than just by traditional demographics, such as age and socio-economic characteristics, and historic sources of information. For example, only a small percentage of this group of travelers is inclined to review the travel section of the newspaper, and they are greatly interested in offerings such as photos or videos of hotels, photos of local surroundings, as well as maps. Given their participation in online networks and blogs, this group is more inclined to seek out recommendations of family and friends than those of a traditional travel agent, and this group also has a great affinity for self-service, including the self check-in and check-out features at hotels. Finally, it is possible that this population segment may be less interested in participating in airline frequent flyer programs.

Given that mobile devices are the choice for communication for Millennials, including communications with airlines, airlines must exploit the full potential of mobile devices in terms of engaging with passengers, improving the customer experience, cutting costs, and increasing revenue. While airlines are already offering the ability to book flights, check-in, board, and obtain real-time flight updates via mobile devices, PDAs and web-enabled mobile phones could also be used for baggage tracking data, biometric data, and payment data, thus making the process truly mobile and paperless. British Airways is already experimenting with the use of the iPhone to give real-time flight arrival and departure information as well as the complete BA timetable, and access to BA's website. This iPhone application is available, free of charge, at the iTunes store. Qantas began to implement a new procedure in 2009 which allows passengers who select their seat online to be automatically checked-in using a barcode that will be sent to their mobile device 24 hours before their flight's departure, and, All Nippon now provides the capability for its domestic passengers to check-in with mobile phones. Air New Zealand introduced the ePass and the mPass. The ePass is a small RFID tag attachable to a mobile phone that passengers can swipe at kiosks, airport lounges, and gates. The mPass provides the capability for web-browsing and mobile barcode boarding.[37]

Airlines can also extend the capability of passengers to re-accommodate themselves in-flight and on the ground (including the ability to track their own baggage), to connect with other modes of transportation, and to connect with their reward systems. To win their loyalty, airlines can develop product features that this generation values, such as continuous connectivity pre-flight, in-flight, and post-flight. Finally, since Millennials are into the "here and now," airlines, as well as all businesses in the service industry, should capture Millennials' feedback right there and then. For example, with respect to airlines, (a) get the opinion from the actual traveler, not a focus group, and (b) get it while the iron is hot—get the reaction, positive or negative, there and then.

It appears that many airlines have neither analyzed intensively the challenges faced by social networking (consider the examples given earlier relating to earlier experiences of jetBlue and United) nor pursued the opportunities that social networking has to offer in depth. In the case of the latter, there are a few examples of airlines that are now embracing the social networking medium including American, jetBlue, and Southwest.[38] Why, one might ask, have the majority of airlines not fully embraced social networking given that (1) it is relatively inexpensive, and (2) it is apparent that certain consumer groups, such as the Millennials, consider it almost a nonnegotiable factor in all parts of their lives, including their consumer behavior? One possible reason for the apprehension is that airlines have more questions than answers when it comes to social media. For example, management wants to have a method of measurement to ensure that any efforts are worthwhile. While it does not cost much, if anything, to become involved with the sites themselves, there is certainly the cost of the employees' time, such as in terms of monitoring the sites. Furthermore, airlines also have the concern that social media does not become a method to solve customer problems. For example, consider Twitter, a social networking medium that involves messaging in real-time. While it is a great tool to convey last minute fare sales due to the immediacy and viral nature of Twitter, and while it would certainly be attractive to a tech-savvy consumer such as the Millennials, is it really an appropriate medium to replace human contact in terms of solving problems? For example, as a manager at Southwest Airlines points out, an airline cannot return a missing bag in 140 characters. Customers must be assured that at the other end of the phone or email there are employees who are trained to address such issues.[39]

Airlines need to delve into the social media world, and determine which tools work for their needs. For example, in terms of providing an outstanding customer experience, Twitter may be an appropriate tool given that its search engine has the ability to track people with respect to their smallest tastes. It is, thus, probably not surprising that hotels have been heavily pursuing the Twitter tool. Some recognize that social media can be used to connect with customers and show that a company really cares, an area in which most passengers would say the airlines are lacking.[40] Social networking can also be used to respond to both positive and negative communications from customers. Take, for example, the case of United Airlines breaking the guitar of a passenger. In July 2009, a musician became an Internet sensation by making a video (and placing it on YouTube) with a song on how United Airlines broke his guitar. Millions of viewers saw the video as it was reported widely in the media. Although United claims to have used the video internally to improve its baggage handling processes, it is possible that United could have made an equally eye-catching video (say, the Chairman playing a guitar) to respond to the negative communication and salvaging the situation.

As for the current economic downturn, the impact of this recession has been deep and wide. First, leaving aside the decline in Economy Class travel, the decline in premium travel is particularly worrisome in terms of its impact on revenue. See Figure 1.2 in Chapter 1. Second, while airlines identify new sources of revenue (ancillary revenue through unbundling the product), passengers are becoming more irritated not only by the number of additional charges, but also, in some cases, at the level of additional charges. In US domestic markets, some passengers have been charged a fee for a change in a reservation higher than the original fare. Here are two examples: a round-trip ticket purchased for US$159, a change in the reservation for the return portion, US$300; a round-trip ticket purchased for US$195, a change in the reservation for outbound leg, US$260. At least for many of those airlines operating within the US domestic markets, such practices, while generating significant amounts of ancillary revenue, are damaging brand equity.

WestJet Airlines, a low-cost carrier in Canada, has a unique philosophy in terms of the customer experience, specifically with respect to products offered. WestJet refers to their passengers as guests, and are very careful not to implement money-saving changes that negatively affect the guest experience. For example, drinking water was originally included in the WestJet ticket price. Instead of taking that free item away, which would most likely have created negative feedback, instead, the airline chose to offer poured water for free, but charge for bottled water. Therefore, the guest has a choice: take the poured water for free, or pay for the bottled water. By providing a choice rather than making the choice for the guest, the airline has achieved greater customer acceptance. In a time when many airlines are "nickel and diming" passengers, WestJet distinguishes itself by its "Care-antee;" a list of items that are included in the price of the ticket for guests, much like Southwest does. This Care-antee lists a myriad of guarantees, such as that the airline will not charge a guest for two checked bags, the airline will not overbook flights, and that the airline will accommodate a guest if her flight is delayed, even if the delay is due to Mother Nature. It has been reported that 92 percent of guests who fly with WestJet will fly them again because of the guest experience.[41]

Virgin America is another airline that is focusing on the experience factor. For example, Virgin America has implemented "Red," the airline's in-flight entertainment system. Red does everything from providing the in-flight safety video, to Wi-Fi access, to movies for purchase, as well as TV shows, music, and games. The system also allows passengers to browse and select meals, snacks, and beverages directly from their seats whenever they wish. A passenger notes the numerous positive aspects, which include friendly staff, dynamic music and 'mood' lighting in the cabins, spacious leather seats, and low fares.[42] Virgin America appears to be living up to its slogan, "This is How to Fly," and is certainly giving the US carriers some serious competition.

Adrian Slywotzky, author of an article in the *Oliver Wyman Journal* states that the name of the game is re-segmenting. Specifically, he calls for the continuous process of re-examining and re-identifying:

- Who your customers are?
- What they want?
- How willing they are to pay?
- How many are there?

Slywotzky recognizes that new customer segments have become a permanent way of life for businesses. Therefore, he urges businesses to consider a proactive approach in designing their companies, knowing that, for example, in the next five years, five new customer segments could emerge.[43]

In this environment, one thing is for sure: value is most important to consumers. What can an airline do in this environment? Henry Harteveldt, of Forrester Research, suggests four pillars that airlines should be using in order to foster successful relationships with their passengers:

- The first pillar is merchandising. Merchandising encompasses a number of factors. First, this pillar includes offering well written descriptions of products and services, as well as extensive visual content whenever possible. This could include clear details of fares, as well as descriptions and illustrations of why your offering is better (lounges, seats, and in-flight entertainment). This pillar also includes truly understanding consumer behavior in terms of cross-selling. Instead of just offering a complementary product to be purchased, such as suggesting a belt with a pair of pants in the retail world, really cross sell the right products to the right customer. Southwest Airlines does an excellent job of this through its Business Select program. Specifically, the airline upsells a customer with what they really need—priority boarding, the ability to make same-day flight changes at no additional cost, and the ability to get a refund or reuse the ticket. These are relevant offerings that provide value to business passengers, especially due to the flexibility factor.

- The second pillar is context. Context gets away from the price factor, and rather hones in on a traveler's needs. For example, InsideTrip.com offers a trip quality index for travelers. Specifically, the site allows the traveler to rate her trip according to factors such as price, speed, comfort, and ease, thus assisting the traveler in sorting through the myriad of options available in the marketplace.

- Engage is the third pillar. Airlines need to engage their customers. Period. The social technology tool, Twitter, which was previously mentioned in the Millennial section, has proven to be a successful means of doing so. Consider the statistic that there are more people on Twitter than live in New York City. Companies are using the social technology tool as a means of conveying special offers only to those who are Twitter users. For example, the Marriott hotel in Napa Valley posted a special offering to only those on Twitter, and it sold out. It is important to note that a few airlines have already embraced Twitter in a similar manner. Specifically, it has been reported that jetBlue and United Airlines have both used Twitter to offer exclusive promotions; these special fare promotions are referred to as "cheeps" by jetBlue and "twares" by United.[44]

- The final pillar is value. As many have pointed out, value is key in this environment. Value does not necessarily mean price. What if airlines let their passengers create their own value? For example, studies indicate that 43 percent would pay for a guaranteed spot in an overhead bin, while

> 38 percent would pay for a guaranteed seat away from families traveling with babies, and finally, 46 percent would pay for an insurance policy that guaranteed the airline would take care of them, regardless of the circumstances, in the case of a delay or cancellation. These examples illustrate that the airlines have assets that travelers value and would, therefore, pay for, if they are not already receiving the benefit through their loyalty program.[45] For more information on value, see Chapter 5.

Expedia and Travelzoo are two examples of companies in the travel industry that recognize that consumers want value, and that value information does not necessarily mean price. For example, Dara Khosrowshahi, CEO and President of Expedia, reported that of consumers who are searching for airfares on Expedia, only 17 percent bought based on the lowest fare offered. However, 83 percent bought based on a higher value, because they also value other factors over just the lowest price, such as the number of stops, and flight times. Expedia is working to provide more value on the air travel side of its business by offering more content to its consumers. Specifically, Expedia bought the company, SeatGuru.com, in order to have more value-added content. SeatGuru.com provides a seating chart of the plane for the passenger, and indicates which seats are superior and which are substandard through a color-coding system. Expedia also launched a fee estimator application to provide value to the consumer. The objective of this application is to make the offering a positive one. If a travel provider is going to charge for an amenity, it is important to be up front about it, rather than hiding the charge which creates a negative experience as the consumer feels cheated. Khosrowshahi stresses that it is imperative to build trust with the consumer by providing information. He gives the example of Amazon.com, a company which has built trust with its customers by providing information.[46]

Likewise, Travelzoo provides numerous examples of how a travel company can supply value for the user. Being a publisher of best travel deals, it conducts comprehensive "research on the quality and competitiveness of each offer," "evaluates the allure of an offer," and has its "test booking center test-book each offer to ensure a smooth user experience." In another example of value, Travelzoo's subsidiary, fly.com, not only provides Economy, but also First Class fares on a researched route so that the user can see if there are cases where the difference between the Economy Class and First Class fare is small enough to upgrade. Next, there is a button through which a user can get an answer to the question of why she should select this particular airline. This last feature provides value to the airline by enabling the airline to not only differentiate its product in areas relevant to the user, but also be able to upsell.[47]

Continental Airlines is an example of a company that recognizes that consumers want to be engaged. The airline has a department that is solely dedicated to customer experience, and one of the roles of the department is to be the voice of the customer. Continental recognizes the need to understand the patterns contained in consumer feedback, and is committed to doing something about it. One means of doing so is by participating in the FlyerTalk.com (FT) conversation. FlyerTalk.com is an online community that has been set up by frequent travelers in which passengers can share their views, as discussed in the previous book.[48] Continental recognizes that the quality of the feedback that the airline receives from FT is astounding, and, in many respects, it is better than any other customer feedback channel. The feedback is detailed, tactical, and is in real-time. The airline recognizes the importance of admitting when they have done something wrong, and takes action, as appropriate. Furthermore, the airline indicates that changes are made almost weekly (some large, some small) as a direct result of FT, but it is not just the Customer Experience department at Continental Airlines that is involved in FT. Many of the airline's employees frequent FT on a regular basis, all the way up to the CEO.

This dialogue is not just limited to the online community. Continental Airlines has also hosted four gatherings in Houston, Texas. The gatherings actually started out as a bet between CEO Larry Kellner and a FlyerTalk participant in which Kellner agreed to host a weekend gathering if at least 100 FlyerTalkers would be willing to come to Houston, at their own expense, to meet with Continental. Nearly 200 FlyerTalkers showed up that year, followed by nearly 300 the following year, nearly 400 the next, and most recently, in February 2009, the event had to be capped at 525 attendees. Attendees have to pay their own way to attend the gatherings, while approximately 100 employees from several departments volunteer on their own time to assist with designing, developing, and hosting this event, enabling the airline to manage the cost factor. Over the years, these gatherings have proven to be a unique opportunity for FlyerTalkers to interact with Continental's management team. Most importantly, these events have provided an appreciation for what goes into running the airline day after day. These events are also great for the airline, as employees have the opportunity to meet customers who are as passionate about Continental as themselves. Like with the online forum, the airline has made a number of changes to its product as a direct result of these events.

First, a passenger will now have access to virtually the same information that the airline has in its website or at its call center. So, instead of passengers asking why they cannot be upgraded or where they are on the list of upgrades, passengers can get access to that information on their personal electronic devices (PDAs or mobile phones). Second, if passengers see that the flight is listed as being on time on the monitor at the gate, but the plane is not physically there, they are able to get more specific and up-to-date information about the flight on their mobile phones. In fact, passengers can actually track the status of their airplanes a number of segments back. Third, some passengers were concerned that even though they have achieved elite frequent flyer status, they would lose that status after they retire and travel less. Continental devised a system which enables a category of passengers that have accumulated a certain number of miles to be able to keep their status indefinitely. The point is that not only are progressive and proactive airlines listening to their customers, but that they are changing their products and services to respond to their customers' needs to improve the customer experience.

As for value, Air New Zealand is focusing on the "airport experience" of its passengers traveling in domestic markets. The airline has transformed its processes to make it really simple to get on an airplane. The pre-departure time for domestic passengers at airports within New Zealand has been reduced to 15 minutes.[49]

Employee Capabilities

Great customized consumer experiences require great employees. Employees who are unsatisfied, ill trained, and unengaged are not going to be able or be motivated to provide outstanding experiences for customers. Yet many airlines think this is going to magically happen. It is ironic that the frontline employees at some airlines are often those with the least experience, the lowest pay, and the most ill-equipped within the company. Yet they are the ones charged with one of the most important jobs, orchestrating a great customer experience for passengers. In terms of service calls, one study indicates that customers are most concerned with two areas when contacting a business: (1) if the frontline employee is knowledgeable, and (2) if the problem is resolved within one phone call. However, it has been reported that often these factors are not even on customer service managers' radar screens. Instead, often these managers are more concerned with metrics such as time on hold and minutes per call.[50] There is a certain hypocrisy between what an airline espouses and what the airline provides to the customer. The disconnect between the customers and the ill-equipped employees is evident in any number of areas. As one executive asked: Why does the operations department put the most unsympathetic employees on the frontline in the lost baggage department? The passenger is already frustrated and upset, and then is faced by an unsympathetic employee who merely hands the passenger a form to fill out, thus worsening the already poor experience?[51] The ability to deliver good experiences involves an individual's service aptitude, and these soft skills must be learned through training conducted by the airline. Specifically, service aptitude refers to the ability of an individual to recognize and act upon opportunities to exceed a customer's expectations. But how many airlines take the time to teach such soft skills?

Research indicates that, on average for companies in general, a company focuses more than 90 percent of its training on hard skills, including operational skills, product knowledge, and technical skills. This leaves less than 10 percent for training of soft skills, including customer service training, experiential training, relationship building, role playing, and service recovery training.[52] Knowledgeable airline executives point out that the statistics are worse within the airline industry. Even if one assumes that airlines are typical, how can an airline achieve the goal of providing phenomenal consumer experiences if it spends less than 10 percent of its employee training time on delivering such experiences?

According to author John DiJulius, insufficient training is one of the largest drivers of inconsistency and lack of great customer experiences. Within the airline industry, many minimize training not only due to costs, but also due to the fact that they are already so far behind and/or short-staffed, that when the employee comes on board, they throw the employee into the workplace without the proper training. This is a recipe for disaster for both the employee and for the passenger. Passengers do not generally care if the employee is new or not; they want their needs met efficiently, and rightfully so. Service should be consistent and seamless for each and every customer. At the same time, how can a new employee, thrust in front of customers with little to no training, provide an outstanding customer experience? The airline is setting both the employee and the customer up for a poor experience. While this seems rather obvious, unfortunately, this situation occurs far too often, but could be easily remedied if airlines would take the time, effort, and commit the resources to conduct the proper training. The Container Store, based in Texas, recognizes the need for extensive training for newly hired employees. Specifically, the company has all new employees go through 241 hours of training during the first year of employment. That is a far cry from the retail industry average of just eight hours! Furthermore, the company invests in a full-time sales trainer for each of the company's stores. Such investments have paid off; employee turnover has remained well below the average for the retail industry. It is not surprising that The Container Store has ranked near the top of *Fortune*'s list of 100 Best Companies to Work for numerous years.[53]

If an airline wants to augment the experience for its passengers, then it needs to start with engaging its employees. Invest in employees so they will, in turn, invest in customers. Author Shep Hyken recognizes that only companies who go above and beyond for their employees can expect their employees, in turn, to go above and beyond for the customers. He notes that the manner in which a company treats its employees is a directive to them on how to treat customers, and this refers to all employees, administrative, management, and technological, not just frontline employees. It often requires the help of employees who are not directly interacting with consumers to actually accomplish what the customer needs done. Therefore, all employees must excel at delivering outstanding customer experiences. Otherwise, companies actually end up frustrating and upsetting its customers due to a lack of cooperation of employees internally. Customers want to complete their transaction, whether it be a purchase or a service; they do not want to be passed on from division to division, nor do they want to be told how the company they are patronizing is organized, operates, and so forth. Operational Dynamics, an engineering consultancy group headquartered in Sydney, Australia, recognizes the need for interdepartmental communication in order to provide quality customer experiences. Specifically, the company notes that most organizations run into trouble at the boundary between business units or departments, but, in reality, many customers' needs require communication to occur among two or more departments.[54]

Interestingly, it can actually be the mentality of airline management that is sometimes the underlying factor behind why an airline is unable to deliver reasonable service, let alone superior experiences. Specifically, management often think that their situation and operation are unique and, in turn, use this as the excuse why they cannot provide better service, let alone experiences, for passengers. Examples include managements' lack of control over weather, mechanical failures, and air traffic control at major airports. Another example is the excuse that an airline cannot be choosy in terms of candidate selection given that it is only able to offer low pay. If such an excuse were valid, then how could service industry leaders, like Starbucks, Disney, Nordstrom, and Southwest Airlines, who do not pay more than the industry standard, attract and retain such consumer-centric employees? Consider Disney. They select and hire their employees from the same labor pool that other companies have access to, and the company offers the "going rate" in terms of compensation. The reason that Disney's employees shine in terms of providing outstanding customer experiences is that the company equips them with the capabilities to do so.[55] The transformation occurs during the training process. In the airline industry, examples would include Southwest and Singapore Airlines. Most airlines totally miss this opportunity.

While management's leadership and employee training are key factors in delivering excellent customer experiences, it can also be the frontline employees, themselves, that keep an airline from delivering superior consumer experiences. Consider DiJulius's statement: "Customer loyalty is won or lost at the front lines of each individual location."[56] First, many employees misunderstand the whole premise behind the customer experience.[57] Some employees cannot distinguish between the transaction and the experience. They think that if they merely do their job, such as a cashier ringing up a customer's purchase quickly and accurately, then the employee has given the customer a great experience. While efficiency and accuracy are certainly key attributes contributing to the experience, there is so much more to it. A customer is more than a transaction; a human element is needed. The British sandwich shop, Pret A Manger, recognizes that they are selling much more than just sandwiches. Rather, they are selling a whole platform, which includes a focus on the customer at every touch point.[58] An employee's attitude is a key factor. Authors Larry Selden and Geoffrey Colvin believe that in order for a company to be truly customer-centric, the company needs to hire employees who are genuinely and fully comfortable with, and believe in, the idea that customers are at the center of everything. Once again, within the global airline industry, Southwest and Singapore Airlines are great examples. Selden and Colvin point out that it is surprising how many employees view the customer as a pain with which they have to deal; thus, the authors adhere to the old rule of thumb of hiring for attitude and training for skills. Southwest Airlines is a living example of this philosophy. Employees must buy into the idea that customers are a positive source for the company.[59] Furthermore, there is also the consideration of how the employee interacts with fellow co-workers and management in front of the customer. For example, discussions about breaks, start/finish times, etcetera should never be held in front of/during an interaction with a customer. This totally affects the customer's experience, and yet most employees have no clue as to the negative impact they are creating. Again, the idea is that the customer is the center of everything.

Let us be sure that there is no confusion between the terms customer service and customer experience. The basic component of customer service is the processing of a transaction with accuracy, efficiency, courtesy, and friendliness, as noted above. In the case of the airline business, it means the ability to make reservations with ease (through a user friendly website or call center), a hassle-free check-in and boarding at the airport, courteous and friendly service on board the aircraft, an on-time performance, and efficient baggage delivery. These are nothing more than what a customer would expect as customer service. The concept of customer experience goes much further. Customer experience includes additional attributes, such as various levels of personalization. The following are some examples of initiatives being considered by truly customer experience-focused airlines.

- When a passenger makes a reservation on an airline's website for a trip that involves a tight, but legal, connection, the screen could provide a clear and loud alert of the possibility that the passenger might miss the connecting flight given the conditions at a particular airport. This feature may cause the passenger to select a different inbound flight if that flight is connecting to an intercontinental flight that operates only once per day, and, worse, if it is only four times per week.

- An airline could have a cart waiting at the gate for a couple arriving on a flight that is late so that the cart could enable the passengers to make the connection, regardless of the level of status of the passenger in the frequent flyer category. With such a customer experience, perhaps these passengers would make this airline their preferred carrier.

- A flight attendant could go to each and every passenger on a delayed inbound flight to inform them, individually, about the status of their connecting flight, and what options the airline offers to those passengers who will miss the connection for sure.

- There could be a group that tracks passengers who have left valuable items on the aircraft (such as laptop computers containing vital information for business travelers), as opposed to having the items sent to the lost and found department.

- An airline could try to contact passengers before they arrive at an airport when a flight is delayed an exceptionally long time or when it is cancelled, each and every passenger, not just the frequent flyers (or even worse, only those in the high mileage category).

- An airline could allow a passenger, even on a full flight, whose seat breaks (the back will not recline) to move to the premium cabin, if a seat is available.

For contrast, next are some examples of situations that meet the criteria of adequate customer service, but do not elevate the concept of customer service to customer experience. Again, a few airlines are working to improve the situation.

- A four-hour, nonstop flight in a US domestic market departs and arrives on time. It departs at 8 AM. Even though the flight attendants are friendly, there is no food available, even in the First Class cabin, free or for sale). Passengers are not given this information prior to boarding. The return flight at 5 PM has the same characteristic. This time, though, passengers are given the information by the boarding agent, literally as the passengers have begun the boarding process.

- Passengers on a delayed inbound flight are informed by a flight attendant (in a very cheerful tone) that the connecting passengers need not worry since they will be met at the gate by an airline agent. There is an agent at the arriving gate, but only one agent who must deal with 40 passengers with different connections.

- A flight on a regional feeder carrier is about to depart. The weight and balance information requires that four bags be removed. The airline thinks that it is going beyond the normal customer service level by informing the affected passengers during the flight that their bags were removed because the aircraft was too heavy to take off. How does that enhance the customer experience of those passengers who then go on to board an intercontinental, connecting flight, and arrive at their destination without their bags? Even if the luggage is delivered to their hotel 36 hours later, the experience has been negatively tainted. According to the airline, it went "out of its way" to not only get the information to the passengers while en route, but to also actually deliver the bags to their final destination rather than make the passengers return to the airport. The fact that these were business passengers who had to make important business presentations using the equipment in their bags is not a concern to the airline.

The difference between standard customer service and customer experience (a fundamental challenge for airlines), is the ability to meet the customers' needs through interaction to personalize the service as much as possible. In the case of airlines, challenges come from the fact that (a) they handle large volumes, may that be 150 passengers on a flight or 50,000 passengers during a day at a large airport, (b) they are dependent on external factors, such as weather and airport congestion, and (c) there is very wide variation in the fares paid by passengers on the same airplane. In the last case, every passenger has the same expectation (a good customer experience) whether the passenger paid US$500 for a ticket in Economy Class, or purchased a US$5,000 ticket for travel in Business Class.

The personalization aspect is intrinsic and integrated in everything that the airline does. In the first case, it cannot be quantified, measured, or described specifically in manuals. It is also more than just a smile or calling a customer by name. In the case of airlines, it is making an emotional connection with a customer, and offering a customer a solution to a problem. If a customer is going to miss a very important business meeting or the ability to fulfill the responsibility of being the best man in a wedding, then saying "sorry," that the customer is going to miss the flight is simply not good enough. Finding a solution to the customer's problem elevates customer service to customer experience. In the second case, customer experience cuts across different functions (operations, marketing, and so forth), exists at every touch point, and, it varies by customer.

One way to appreciate the problems faced by employees is for the very senior members of management (including the CEO) to engage with employees in their individual work environments. Rob Fyfe, the CEO of Air New Zealand, represents an example of such engagements. He is quoted in an interview stating that he spends at least one day a month actually doing different functions within his airline; being a flight attendant, a baggage handler, a call center agent, and a sales person (accompanying a sales person as a trainee sales person). In reverse, he also has employees (such as flight attendants, pilots, and check-in agents) spend a day with him involved with such activities as sitting through meetings or meeting with the Prime Minister.[60] It is this type of engagement on the one side, and the one between senior management and customers on the other side, that can lead to customer-centric products and services as well as processes, systems, and incentives as discussed below.

Executives of top service companies understand the shift from focusing on market share and setting profit goals to focusing on the employee (especially those on the front line) and on the customer. These executives recognize that factors such as recruiting and training practices, performance-based compensation, investment in people, and technology that supports frontline employees drive profitability in the new service arena. Furthermore, they understand what is referred to as "the service-profit chain," as described in a recent article in the *Harvard Business Review*. Specifically, the chain involves the linkage of employee satisfaction and customer satisfaction. For example, elevating the quality of internal service, through such actions as equipping employees with the skills as well as the power to serve customers, boosts employee satisfaction. This, in turn, drives productivity and employee loyalty, ultimately elevating the external service value, thus increasing customer satisfaction and loyalty.

In terms of recruiting the right employees, leading companies recognize that attitude is a key factor. Retailer Nordstrom, famous for its excellent focus on customers, recognizes that attitude is a key factor in selecting candidates, not just the skills listed on their resumes. Specifically, Nordstrom follows a slogan (similar to that of authors Selden and Colvin) that drives home the importance of this concept in the customer experience arena: hire the smile (attitude), train the skill.[61] An executive at Four Seasons hotels also recognizes the link between attitude and outstanding customer experiences by stating that they believe that certain attitudes predispose candidates to being effective deliverers of experiences.[62] Southwest believes that only looking at attributes which can be quantified causes the company to miss out on, what Kelleher referred to as, the heart of the business, people. Southwest Airlines has a unique feature in its recruiting process. Specifically, the airline regularly invites its frequent fliers to participate in the selection of flight attendants. Who better to be involved in choosing the frontline staff than the customers themselves? Finally, the British sandwich shop, Pret A Manger, also recognizes that it needs extraordinary people in order to deliver a great experience. In order to help facilitate this need, the recruiting process includes a unique aspect. Specifically, after passing several interviews, successful candidates are put to work for a day in one of the Pret A Manger shops. Employees are then allowed to vote on the candidate in terms of whether the candidate should be hired. This unique step in the recruiting process helps ensure both a good fit as well as good teamwork.[63]

According to John DiJulius, author of *What's the Secret? To Providing a World-Class Customer Experience*, first and foremost, world class companies view their employees as their primary customer. Such companies take the time to train employees of each internal department that they have customers whether they are on the frontline or not; they make the departments recognize that their customers are the other departments or components within the company. Everyone has customers, whether they be external or internal, and, therefore, everyone has the common goal of pleasing the customer. They even help employees understand what it feels like to be a customer of the company. For example, the final stage of the multiple-step employee orientation at the

Four Seasons hotels involves workers, regardless of whether they are a housekeeper or a front desk clerk, staying at a property, free of charge, with a guest, along with free dining. What is the point of such an exercise? They are learning what it is like to receive outstanding experiences from the reciprocal side.

Such companies instill a sense of meaning in their employees. They are not just reporting to work, but rather are on a mission. This is a key attribute of successful consumer experience companies such as Disney and Ritz-Carlton.[64] This is also true of the Four Seasons hotel group. The company's founder, Isadore Sharp, created a culture where the desire to exceed a guest's expectation was at the core. Leading by example, Sharp has created a global force of employees focused on consistently doing what is right for the customer. This could be as small as knowing of, and having, the guest's favorite soft drink available in the refrigerator in the guest's room. The employees pay attention to every detail, even if it may seem insignificant. It is a culture that promotes and requires the engagement of every employee at every level, from hotel managers and concierges to housekeepers and doormen. Most importantly, however, is Sharp's rule of not implementing savings at the expense of the guest. If employees can find ways to cut costs and save the company money without affecting the customer experience, that is acceptable. However, cost cutting at the expense of a guest is not. For example, making the decision to cancel the pianist playing in the dining room during slow months of the year can affect a customer's experience, advised the founder. What if a couple had saved enough to splurge and stay at the hotel during the slow time? That couple would have been robbed of the total experience.[65]

In terms of satisfying these employees once they are hired, having the right working environment is one of the most important factors, and a big part of this environment is the ability, as well as the authority, of employees to achieve results for customers. If employees are not given the ability, in terms of the proper training and tools, or the power to make customers happy, then a high level of employee satisfaction will not be achieved.

Customer experience-focused companies spell out the exact level of service that is required by its employees. DiJulius refers to this as the "nonnegotiable experiential standards" that everyone in the company must follow. For example, when Disney hires new employees, or "cast members," as the company refers to them, the nonnegotiable service standards, which include such elements as courtesy, efficiency, and safety, are clearly defined to potential candidates. This helps ensure that there is a proper fit of the potential employee with the Disney culture. Such companies provide and review training with employees regularly. This practice not only refreshes and reinforces policies and practices in the minds of employees, but it promotes consistency as older employees are hearing the same message alongside new employees.

Customer experience-focused companies engage different departments of the organization in discussions in order to allow employees to have the opportunity to share their direct experiences from the frontline on how the company can deliver better experiences to its customers. This practice not only engages employees in decision making, but also helps to obtain direct buy-in from the employees who are actually going to deliver the new initiatives, rather than just having mandates issued from management. For example, David Neeleman, former CEO of jetBlue Airways and current founder of Brazil's new Azul Airlines, recognizes that a company's call center employees are the key link to the customers. Specifically, Neeleman reminds the company's executives to consistently check in with the company's call center agents, as while management thinks they know what is going on, it is the frontline agents that really know firsthand what is happening with the company's customers.[66] Such companies strive to teach their employees to not become too task focused, but rather to focus on the bigger picture of the customer experience. Finally, such companies are always striving for an even higher level of excellence, despite the fact that they are already leaders. For example, The Ritz-Carlton, Sarasota, which had already been ranked the third highest in terms of guest satisfaction ratings of any Ritz-Carlton, was pursuing opportunities in terms of what more they could do so that they could become the highest ranked property.[67]

Outstanding service companies employ people who are able to deliver excellent experiences in good times, but are also able to rectify issues in bad times. For example, Southwest Airlines gives the power to its employees to turn a customer's problem into a positive experience. Specifically, the company allows its frontline employees to do whatever they feel comfortable doing in order to achieve customer satisfaction. How many airlines maintain such a policy? Similarly, Xerox gives its frontline employees the power to replace up to US$250,000 in equipment if needed to achieve customer satisfaction.[68]

These leaders are in tune with what their customers value. As another example, Southwest ranks very high in terms of customers' perception of value, yet the airline does not offer some of the features its legacy competitors do, such as seat assignments, meals, or link its reservations systems to other airlines. Then how does Southwest achieve such a high ranking? The airline consistently offers low fares, on-time service, and most importantly, friendly employees; all factors upon which its customers place a high value. How do they know? Southwest has processes and systems in place so that the airline is in consistent communication with its employees who report the findings of their daily interactions with customers up to management.[69] Furthermore, great customer-oriented companies, like Southwest, make the effort to have management spend time observing employees and customers. For example, Herb Kelleher, former CEO of Southwest Airlines, could be found frequenting the places where his employees and customers were located (in the airports, on the planes, on tarmacs, etcetera). Successful leaders spend a great deal of time observing, listening, and considering suggestions for improvement in terms of processes and systems. Such actions drive both employee satisfaction as well as customer satisfaction.

Finally, companies who provide a "world class" customer experience are "world class" in all respects; to its employees, to its customers, to its community. In order to become a "world class," customer-focused organization, a company must be recognized as an employer of choice, which according to DiJulius, requires four main characteristics: providing a great place to work, providing great training, having superior customer service, and offering unlimited opportunity for its employees.[70] The last attribute is where many companies fail; companies must invest in their employees, help them grow, and bring out their winning qualities. In order for the company to reap the benefits, it must also give.

If a company does not offer anything to the employee in terms of development opportunities and a potential career path, then that employee will most likely leave sooner than later. Leading companies create an internal culture that promotes attracting, hiring, and retaining the people who are capable of promoting the common goal of "world class" experiences. "World class" companies measure customer satisfaction, tie their employees' performance and incentives to customer satisfaction, and celebrate and reward those who are successful at delivering and achieving customer satisfaction. How many airlines fit this mold?

Processes, Systems and Incentives

To provide a good customer experience, employees need to be not only selected carefully and well-trained, but they also need to be well-equipped with customer-focused processes and customer-friendly systems. Additionally, airlines need to empower, to some degree, their employees to appropriately use the systems to create outstanding experiences for passengers. With the exception of a few US airlines, this is not the case for the majority of US airlines. With respect to the majority of US carriers, the employees who actually interact with the passengers are poorly trained, are poorly equipped with the tools to take care of customers, and are handcuffed by the rules, practices and systems that do exist. In summary, they provide quite the opposite of outstanding customer service, let alone customer experiences.

In some cases, airlines do not even use the processes and systems that they do already have in place to their full potential. For example, consider customer relationship management systems. Many airlines already have them in place, but they do not use them as a tool to personalize interactions with the customer and to ultimately provide outstanding customer experiences. On the other hand, some companies do recognize the power of such systems. Consider the Mandarin Oriental hotel group, for example. The company has its customers complete a questionnaire prior to arrival, including such specifics as time of arrival, etcetera, and then the staff utilizes the information to customize the experience for each guest. Furthermore, the company also leverages technology in order to augment and personalize its offering. For example, if a room service order is not completed and delivered correctly, a memo gets transmitted to all staff over the hotel's rapid response communication system, enabling employees to rectify the situation in a timely manner.[71] Another example is the Irish grocer, Superquinn, which has a loyalty program, as most grocers these days do. What makes Superquinn stand out is how the company uses its program to create services that are valued by the patrons. Specifically, shoppers can actually obtain reward points for pointing out problem areas in the store, such as an out-of-date product or shopping carts that are not working properly. What an opportunity for airlines! This unique use of a loyalty program has created a fun and rewarding way for shoppers to also serve as quality control agents for the company. Furthermore, the company uses its loyalty program to personalize the customer's shopping experience. Specifically, each time a customer swipes her loyalty program card when checking out, her name appears on the cashier's screen, allowing the employee to address the customer by name.[72]

In terms of loyalty programs, airlines need to revamp their systems. A recent article in the *Association of National Advertisers* (ANA) magazine recognizes that points and prizes are no longer enough to attract, reward, and retain customers. Specifically, the article illustrates that while "points-for-prizes" programs are still popular, the new trend for customer loyalty programs extends far beyond to include everything from enhanced customer experience to social networking opportunities. For example, American Airlines' program promotes something different: an experience. Specifically, the goal of American's program is "to expedite and smooth our customers' journey," states a managing director of customer experience for the airline. One loyalty marketing consultancy firm recognizes that this new trend of "soft" benefits will be the key differentiator for successful loyalty programs in the future. Successful and effective loyalty programs require a commitment to constant improvement and relevance assurance. As one loyalty consultant notes, being a member of a loyalty program does not necessarily denote engagement with your company. Loyalty programs must be assessed on a regular basis.[73]

On the other side of the coin, some airlines recognize that service defects can and do arise, and they proactively have processes and systems in place to address them. Moreover, these airlines take the time to teach their employees how to use these processes and systems to deal with such situations so that their customer leaves focusing more on the great experience she encountered in terms of rectifying the problem, than on the problem itself. For example, Disney is proactive about the fact that a number of families at the end of each night will forget where they parked their car after a long day at the amusement park. While this is not Disney's fault, nor do they assess blame, they have taken steps to address the recovering problem. The company has implemented a process in which it keeps track of which sections of the parking lot are being filled at different times of the day, so they simply ask if the guest remembers approximately when they arrived that day, the employee looks up what area was being parked in at that time, and then they drive the guest to that section.

When something goes wrong at a Ritz-Carlton property, its employees are empowered with the authority to do whatever it takes to correct the situation, and, in turn, make it into a positive customer experience. For example, a couple were on their honeymoon at one of the Ritz-Carlton properties, and their luggage had been misplaced during their travel. They literally had nothing to wear, but their wedding clothes. The employees turned the mishap into an opportunity, and moved into "service recovery" mode. The staff opened up the gift shop, even though it was after hours, so that the couple could choose some clothes, toothbrushes, etcetera, all complimentary. The staff was extremely apologetic, and sincerely cared about the situation. The bags were eventually found, but in the meantime, the staff was able to turn a miserable experience into a positive one using their attitude, ability, and authority to think out of the box, coupled with the resources at their disposal to rectify the situation[74] (for more on Ritz-Carlton, see the case studies at the end of this chapter). This is certainly not usually the case when the airlines lose baggage. Also consider the case of passengers having to sit in an airplane on the taxiway or ramp. While most airlines take an "Oh well" attitude and consider the problem to be out of the airline's control, Southwest Airlines takes a different approach to this service defect. For example, pilots walk through the cabin offering updates and answering questions, flight attendants provide information regarding connecting flights, and furthermore, the airline has been known to even send apology letters to the passengers, including complimentary ticket vouchers. The airline demonstrates to the customers that the company cares about the passengers through these actions. Not many airlines show this level of concern for their customers.

Ritz-Carlton has gone so far as to implement a system to check the company's processes and systems. Specifically, the company has implemented what they refer to as "Day 21." This program allows new employees to openly discuss the positives and negatives they have encountered during their first three weeks at the company. For example, in this forum, new employees may discuss whether they feel that they have been equipped with all the tools they've needed to achieve success, the quality of those conducting the training, and whether the culture described during orientation is in fact what the new employees have experienced thus far. Ritz-Carlton takes the time to conduct such a program in order to be proactive and have the chance to "make things right" before employees become disillusioned or disengaged.[75] This appears to be an extremely vital exercise given that research conducted by the Corporate Executive Board indicates that a shocking 75 percent of newly hired employees at companies across America feel that their employers are not living up to what was promised.[76] By implementing such a process, however, it is evident that Ritz-Carlton truly listens to its employees and shows the employees that the company cares (for more on Ritz-Carlton, see the case studies at the end of this chapter).

In general, airlines do not take sufficient time to implement the correct processes and systems, let alone processes and systems to validate to the original ones. A few progressive airlines are beginning to implement processes and systems in order for their employees to have the appropriate infrastructure to serve customers, and, these airlines are continually monitoring and reinforcing these practices. Employees' performance reviews, as well as incentive systems are being tied to such practices. One example of a US carrier that is working toward implementing such practices is American Airlines. Specifically, the airline has implemented a program in which it rewards its employees for behaviors that augment the overall customer experience. Under this incentive plan, monthly targets are set for American's customer teams, in which employees can earn up to US$100 per month, or an additional US$1200 per year, for achieving customer satisfaction and on-time performance goals. Customer satisfaction is defined as the customer's experience at touch points, including at the airport and on board the plane, as well as the likelihood of the customer to recommend the airline to others. The plan is crafted to ensure that it measures those factors that are most important to customers, such as baggage handling and courtesy.[77]

It is important to remember the power of the "live person" element in customer experience. While the Internet is an amazing tool that can be very useful and efficient, a live person, if equipped with the proper capabilities, in certain circumstances can truly augment the customer experience. How many times has a customer had an issue that she really wants to discuss with a live person, only to be directed to a website or even worse, put through a number of options when trying to call and reach a live person to help resolve an issue? As Jim Garrity, retired CMO of Wachovia Corporation recognizes, the employees and the actual brick-and-mortar structure (in this case of a financial center) can be incredible assets in constructing the customer experience.[78]

The last section of this chapter includes three case studies involving three different industries: the restaurant industry (In-N-Out Burger), the hotel industry (Ritz-Carlton), and the online retail industry (Zappos). These businesses are different from the airlines, especially in terms of being affected by external factors such as air traffic control, mechanical, and weather issues, but also because airlines are a means of getting to a destination, whereas the other businesses are the destination. However, that being said, these businesses do offer insights into new ways to conduct business in areas that are more similar to airlines, especially in terms of leadership, training and motivating employees, and fostering a corporate culture conducive to augmenting the customer experience.

Case Studies in Consumer Experience

In-N-Out Burger

In-N-Out Burger is a family run fast food chain that operates in the western United States. Despite its small size and family nature, it is able to survive in one of the most competitive industries. McDonalds, in particular, which has been noted as the Wal-Mart of the fast food industry, poses a daunting challenge. The one difference that makes this fast food chain stand out from all the others is its devotion to its unique customer experience.

In-N-Out Burger has several distinct characteristics that differentiate it from all other fast food offerings. First, In-N-Out Burger is built on an ultra simple premise. The company offers four basic items on its menu; hamburgers, cheeseburgers, fries, and shakes. A side menu offers coffee, milk, and soda. It started that way at its first store, founded by a family in a suburb of Los Angeles, California in 1948, and it has remained the same over a half of a century. The co-founder, Harry Snyder's motto was to (a) keep it real simple, and (b) do one thing and do it the best you can.[79] This practice is in direct contrast with the actions of its industry competitors, which are constantly adding new offerings (and thus moving away from their core business) in an attempt to stimulate business. In-N-Out Burger has refrained from diversifying into other main offerings such as chicken, fish, salads, and soups, or extras such as sundaes or cookies; the company has stood by its core offering of burgers. This is not the case with McDonalds, which has added nearly forty items to its core menu since 1955, in addition to a myriad of promotional/limited time offerings.[80]

Second, all of In-N-Out Burger's offerings are made from fresh ingredients. Specifically, the company possesses its own meat processing plant in which its own butchers prepare the beef for the burgers, the potatoes are cut each day in each store for the fries, the tomatoes are delivered fresh every other day and are hand cut in the store, etcetera. No heat lamps or microwaves are used in any of the stores. Interestingly, while In-N-Out Burger does not offer an extensive menu for its customers, the fresh aspect of everything it does offer allows for customization for its patrons' personal tastes. In fact, there is actually a "secret menu" that fans of the chain have spread via the Web, as well as via word-of-mouth. This menu includes names for famous combinations of the company's offerings, such as "the 2×4" which consists of two patties and four slices of cheese or "the Neapolitan Shake" which consists of combining all shake flavors (chocolate, strawberry, vanilla) into one. Since In-N-Out Burger does virtually no advertising, this "secret menu" provides a fun, easy way to promote the company through buzz generated by its fans.

Third, In-N-Out Burger has a unique corporate culture. First, it is still a family-owned business, just as it started over 50 years ago. Second, the culture is extremely friendly, possibly due to the family atmosphere. Third, employees are paid well above average for the industry. For example, store managers are compensated around US$100,000 per year.[81] In addition to being extremely well paid, employees also enjoy extremely generous fringe benefits. It is not surprising that In-N-Out Burger enjoys relatively far lower employee turnover than average for the industry. Low turnover most likely also adds to the culture in terms of the friendly, family atmosphere as well as to the customer experience in that patrons are served by familiar faces rather than a new person behind the counter every visit.

It is important to note that sustaining In-N-Out Burger's unique customer experience does have some negative aspects. First, customers have to wait longer for their food than at traditional fast food restaurants given that the food is always made fresh. Second, on a larger scale, utilizing only fresh ingredients presents growth limitations. Meat is prepared in the one, company-owned plant and is shipped via refrigerated trucks to the individual stores, limiting operations to just a few states in the western US. If the company were to expand, it would require the creation and implementation of new supply and logistics chains. Despite these downsides, In-N-Out Burger has been able to achieve phenomenal results, soaring way above the industry average in terms of both revenue and profit margins.[82]

Ritz-Carlton

Ritz-Carlton is world renowned for its excellence in the hotel industry. How has Ritz-Carlton achieved such a reputation in this service industry? First, the organization's commitment to providing an exceptional experience for each and every guest, each and every time, and at each and every touch point, rather than just settling for offering an acceptable level of service, distinguishes it from many of the other offerings in the industry. Many hotels, as well as other businesses in other industries, start off on the right foot with the customer in terms of making her feel welcome, showing a genuine interest in her needs, etcetera. However, somewhere during the customer experience, the company loses touch, either through a diminished level of service, a lack of follow through, or some other related breakdown. Ritz-Carlton, however, recognizes that it is imperative to continue to connect with the customer at each and every touch point through the customer experience.

Second, Ritz-Carlton helps its employees understand how to provide an exceptional experience, rather than just acceptable service, by distinguishing between purpose and function. Specifically, a general manager at the Ritz-Carlton in Beijing recognizes that the company's leadership helps the staff to connect with the customers by coaching how to create an exceptional experience in what other hotels see as just as function. For example, leadership teaches the staff that the purpose of a front desk person is not to just check-in a customer, but rather it is an opportunity to create an exceptional experience for that guest during her check-in. The company expects and requires its employees to be more than just proficient in their job function. The leadership at Ritz-Carlton stresses that each employee can, and should, do whatever it takes to create a "wow" experience for a customer.

Next, leadership recognizes that the creation of a "wow" experience often requires two or more staff members to work together and communicate regarding the needs of a customer. For example, a housekeeper at the Ritz-Carlton in Cancun went to a guest's room to turn down the bed. The customer indicated that he had a headache, and just requested a bottle of water.

Much to the surprise of the guest, the concierge called the guest later that evening to check on him and to inquire if he needed a doctor, or if there was anything else that the guest needed. It is quite unusual to encounter staff working together in such a capacity, especially at the housekeeper level. However, at the Ritz-Carlton, the staff continually looks for ways to serve its guests, whether it be through a direct request or by offering to those not necessarily asking for service. A trainer for Ritz-Carlton notes that anticipating the needs of a guest is almost an artistic skill, and one that requires staff to observe guests' actions during their stay. For example, one employee overheard that it was a guest's son's eighth birthday during their stay at a Ritz-Carlton property. The employee worked with the kitchen staff to have special pancakes prepared, complete with candles and matches. The employee also worked with the gift shop staff to assemble a gift bag, including a puzzle and a map. All the goods were delivered to the guests' room, and even though none of it was requested, the guests were delighted at the sentiment. Finally, at the Ritz-Carlton, Central Park, some guests happened to mention, in passing, to the concierge that they were going to see a musical while they were in town. Much to their surprise, when they returned to their room, the soundtrack of that musical was playing, and the concierge told the guests that they may take it home with them as a memento of their stay. All of these examples illustrate how Ritz-Carlton excels at fulfilling unstated wishes creatively, a skill that few competitors even attempt.

In addition to employees working together to create such "wow" customer experiences, it also requires that the proper systems are in place. For example, Ritz-Carlton utilizes cross-training among departments, situation training, and technology (such as earpiece radios), all in an attempt to create seamless communication and provide exceptional service. Ritz-Carlton is one business that recognizes the importance of communication across teams as a direct factor in the ability to deliver "wow" experiences. The commitment of the staff to seek out, collaborate upon, and deliver upon opportunities to enhance a guest's experience is the basis of Ritz-Carlton's philosophy and a key differentiator of the company's offerings from its competitors in the industry. This collaboration does not just occur within the staff of Ritz-Carlton. Third party vendors, such as limo drivers, are also engaged to ensure seamless, extraordinary customer experiences. For example, limo drivers communicate names of guests to the doorman of the hotel property, who then, in turn, relays the information to the front desk staff. These are simple and minimal cost ways to augment the experience of the customer. Addressing customers by name and expecting their arrival certainly adds a personal touch to the customer's experience.[83]

Interesting to note, Ritz-Carlton, like other world class service companies such as Disney, is not paying a premium for their employees. Rather, Ritz-Carlton is offering the same compensation as others in its industry. The difference is not in the pay, but rather in the manner in which Ritz-Carlton selects its people, and then how the company treats its people once they join the team. Furthermore, Ritz-Carlton utilizes its employee orientation as a tool far beyond an arena to complete tasks such as filling out new employee forms, tax paperwork, etcetera. Instead, orientation at Ritz-Carlton involves welcoming new employees as guests by the leadership of the company. The leadership is fully engaged in the orientation experience, covering the history, values, and purpose of the business, and most importantly, illustrating how employees can make a difference for its customers. Unlike most companies, Ritz-Carlton's orientation involves giving new employees an understanding of its service culture, and this is just the beginning. After the full, two-day orientation session, unlike many companies who only offer a few hours of orientation, if even that, Ritz-Carlton continually drives home these concepts through ongoing discussions and examples, such as through its daily lineups, face-to-face communications before work shifts. Unfortunately, in most companies, employees are not exposed to the company vision and values once they join the organization, except for at an annual meeting, perhaps, or through an email or memo. Ritz-Carlton, on the other hand, drives the values home to its employees in person and on a daily basis.

Zappos

Zappos, an online shoe retailer based in Henderson, Nevada, located just outside of Las Vegas, has some unique HR practices that have, in turn, yielded both adoring employees and customers. The business was originally founded in San Francisco, but was moved to the Las Vegas area due to (1) relatively inexpensive real estate, and (2) a plentiful pool of call center workers. Zappos, whose name is derived from the Spanish word for shoe, has been in business for 10 years. The company's current CEO, Tony Hsieh, originally became involved with Zappos as an investor shortly after the company was founded, but eventually became a full time employee in 2000. Under Hsieh's leadership, Zappos' gross merchandise sales have grown from US$1.6M in 2000 to over US$1 billion in 2008.[84] The company first offered solely shoes alone, and, while that is still the core of its business, the company has also branched out into other avenues, including apparel and accessories.

What sets Zappos apart? First, Zappos offers an outstanding value proposition for its customers. The company has an inventory of four million pairs of shoes from which customers may choose. Shipping is fast, as the company's warehouse facilities are strategically located just minutes from the UPS hub in Kentucky, with shuttles constantly going back and forth. Not only is it fast, but the shipping is also free, both on purchases and returns. While Zappos promises four-day delivery, the company often is able to achieve next-day delivery, thus delighting customers by over delivering on its promise. The company even allows customers to return products for one whole year after the purchase date. It is interesting to note that even during tough economic times, Zappos recognizes the importance of its customers, especially its repeat customers. For example, instead of cutting the "extra" of a next-day delivery upgrade to save money during the economic downturn; the company launched a new program to reward its loyal customers. Specifically, this program features an exclusive invitation-only website that not only offers the next-day delivery shipping upgraded service for its loyal customers, but also provides extras such as previews of new merchandise. As of 2009, the company had planned to no longer offer the overnight shipping upgrade to first-time customers, but will use those funds to reward loyal customers. In a time when many companies are actually charging extra for the "extras," some airlines are charging for basic items, such as checking bags, and aisle seats.

Zappos clearly understands the importance of not losing sight of the needs of key customers when making budgetary decisions. Not surprisingly, Zappos was ranked number 7 in *BusinessWeek's* Top 25 Customer Service Champs (March 2009).[85]

Second, Hsieh stresses that Zappos is a service company, that just happens to offer products such as shoes and accessories.[86] He recognizes that customer service is not just a department within the company, but rather it involves the entire company. The entire company, every employee at every level, is concerned with the quality of the customer experience.[87] Many companies do not appear to grasp this concept, as evident by their employees' attitude of "It's not my problem." Truly, a customer should never be sent to "Customer Service" to fix a problem; the employee dealing with the customer should possess the ability and the authority to handle the situation right there and then, instead of passing the customer off to someone else in the company.

Third, the company has implemented a list of 10 commandments which serve as the core values and drivers of employee behavior. However, the company does not just list these values, rather they truly live the values and instill them in their employees right from the beginning. Moreover, they make sure that employees continue to follow the values by basing 50 percent of their performance reviews upon core values and culture fit.[88] The first core value, "Deliver WOW through service," empowers Zappos' employees to think and act out of the box. Employees are encouraged to use their imaginations in deriving customer solutions. For example, if an employee ascertains that a customer is having a bad day during a discussion with them on the phone, the employee can have flowers sent to the customer's home. Employees have also been known to send handwritten thank you notes to customers. While these types of practices may seem extravagant and costly, Chairman, COO, and CFO Alfred Lin supports such acts by employees as he recognizes that three quarters of the company's business is derived from repeat customers. Such practices certainly promote repeat customers. Lin also states that Zappos has always been focused on a long-term approach to business decisions. The management feels so strongly about being customer-oriented that it requires all employees, regardless of their job function, to complete 1,200 hours of customer loyalty training.[89] Not too many companies can claim that level of commitment to its customers.

Zappos' nontraditional policy with respect to its telephone representatives also helps foster an atmosphere of thinking out of the box. Specifically, unlike many traditional call center environments, Zappos' telephone representatives do not work from a script, they are not measured on their call lengths, nor are there sales-based performance goals. Interestingly, Hsieh made the decision not to outsource the Zappos call center as he believes that it is a far too important part of the business to be shipped out.

Another one of the company's core values is fun and weirdness. The HR department conducts its hiring practices in alignment with this core value. For example, cartoon-type applications are used as well as unusual interview questions (example questions: "How weird are you?" and "What is your theme song?"). Why ask such questions? Zappos believes that by encouraging "weirdness," they are in turn encouraging and promoting the qualities of thinking outside of the box and of being innovative.[90] New hires are put through a five-week training program which stresses these core values. During this training period, new hires are actually bribed to leave the company through the offer of a US$2,000 bonus, just to ensure that their heart is really with the company if they choose to forego the bonus and stay. Interestingly, only a few individuals actually take the cash and leave the company. While this practice may seem extremely costly, management recognizes that it is far less expensive to find out sooner, than later, if there is not an appropriate fit between the employee and the company. Hsieh indicates such practices are put into place to ensure that the company hires positive individuals. In turn, Zappos works to create an environment which cultivates and nurtures this positive attribute.

On this note, like Google (for more on Google, see Chapter 4), Zappos tries to provide lots of unusual "extras" for its employees to help them keep a positive attitude. Specifically, Zappos does everything from providing free lunch in its cafeteria to covering the full cost of all employees' health insurance to providing a nap room and a full time life coach. Lin recognizes that an employee cannot provide outstanding customer experiences if she is upset with some aspect in her life. Thus, this is the rationale for costs such as a full time life coach on staff. Hsieh also throws a weekly costume party to help ensure employees maintain the spirit of the "fun and weirdness" core value. The general premise behind the company's unique practices is that if they amaze their employees, the employees will, in turn, be inspired to amaze their customers.

Finally, Zappos clearly embraces both branding as well as social technology. Specifically, Hsieh stresses four attributes that he believes are key drivers in building a brand that matters: (1) vision, (2) repeat customers, (3) transparency, and 4) culture. In terms of social technology, Zappos really puts itself out there. Specifically, the company participates in several different channels, including blogs, Facebook, and Twitter.[91] Hsieh credits much of Zappos' success to the unusual amount of openness he encourages, not only with and among his employees, but also with vendors and even other businesses. This high level of openness, coupled with the unique culture that promotes "wowing" both the employee and the customer, is Hsieh's "secret sauce" for success.[92]

The Zappos culture has caused executives from companies such as Lego and Southwest to travel to the Las Vegas-based company to observe and experience it for themselves. Furthermore, it has been reported that Hsieh was asked to visit the Virgin Group to discuss his strategies. The company even has a "culture book" which contains entries actually written by employees as well as vendors regarding the Zappos company culture in terms of what it is that makes it unique as well as successful.[93] Not surprisingly, Zappos was rated number 23 in *Fortune's* 100 Best Companies to Work For (February 2009). The company's rating made it the highest-ranking newcomer to *Fortune's* list.[94]

In its July 23, 2009 online edition, *The New York Times* reported that Amazon made an offer to acquire Zappos for 10 million of Amazon's shares, valued at nearly US$900 million at the current level. It is interesting to note that Amazon tried to expand its own operations to include the sale of shoes (through low prices and convenience) but could not win the market share from Zappos because of the personalized service provided by Zappos. It is also interesting to note that the deal included an additional amount of both cash and stock for employees to ensure that the employees stayed on board to preserve the company's culture.[95] The online shoe retailer is now an Amazon.com subsidiary.

Takeaways for Airlines

It is clear from the three case studies that leading companies always consider the customer experience first and foremost, regardless of the product that they are offering. They can stick to their core business (as in the case of In-N-Out Burger) or diversify (as in the case of Zappos). The key is the identification and delivery of an outstanding value proposition that must go beyond the price component. Only a few airlines can truly make such a claim. Within the value proposition, keep attributes relevant, unconfusing, rational, and value-adding. Let us go back to In-N-Out Burger. Relevance refers to fresh ingredients. Less confusion and rationality refers to menu options. In the case of value-adding, consider Zappos' practice of providing free shipping for purchases and returns. Most airlines' value propositions do not come close to these examples. In fact, in some cases, they go in the opposite direction. Whereas Zappos provides free shipping in both directions, airlines have begun to charge for many products and services that used to be included in the fare. Then there is the question of under-promising and over- delivering. Zappos promises delivery in about four days, but delivers in much less time. Most airlines do the reverse—over-promise and under-deliver.

Another key takeaway for airlines is that the Japanese experience suggests that, in the long run, quality takes time, energy, and resources. It is an attribute valued by customers and it is difficult to copy by competitors. Most airlines only pay lip service to the customer experience. Of course, there are a few exceptions; Southwest and Continental in the US, Virgin and Lufthansa in Europe, Singapore and Cathay Pacific in Asia, and Air New Zealand and Qantas in Australasia. Within this context, the key resource is employees, selection, training, empowerment, retention, and so forth. In the case of Ritz-Carlton, employees are selected and trained to go far beyond simply completing a transaction satisfactorily and accurately. They are expected to be able to anticipate a guest's needs, then be able to satisfy such needs by themselves or in cooperation with other employees. In the case of most airlines in the US, just consistently fulfilling the first requirement would enhance customer experience by an order of magnitude. The second component depends on corporate culture, which begins with leadership, regardless of the size or type of business. This is evident at a small top end business, such as Ritz-Carlton, or a high volume business, such as Zappos. In the final analysis, it is corporate culture that has made Southwest and Continental what they are, and is the one element of the business that competitors have not been able to copy successfully. The corporate culture must now expand its sphere of influence to improve customer experience based on the best practices within the framework of merchandising in other businesses, as discussed in Chapter 5.

Finally, it is important to keep in mind that truly customer-experience companies do not let customer experience deteriorate during financially difficult times. In a recent survey by J.D. Power and Associates, the Four Seasons Hotels and Resorts ranked Number 1 in guest satisfaction among the luxury hotels, receiving the highest score of any hotels in any category (despite making cost reductions during the recession). Even in the economy/budget category of hotels, Microtel Inns & Suites came in Number 1.[96]

Notes

1 To avoid any confusion, the terms consumer and customer are used interchangeably in this chapter.

2 *Unisys Scorecard*, Volume 6, Issue 5, February 2008, p. 4 (originally from Bain & Company, The Gilford Group Limited, November 2006).

3 Based upon "The 6 Laws of Customer Experience: The Fundamental Truths that Define How Organizations Treat Customers," by Bruce Temkin of Forrester Research and DiJulius, John R. III, *What's the Secret? To Providing a World-Class Customer Experience* (Hoboken, NJ: John Wiley, 2008).

4 Griffin, Jill, *Taming the Search-and-Switch Customer: Earning Customer Loyalty in a Compulsion-to-Compare World* (San Francisco, CA: Jossey-Bass, 2009).

5 Bayer, Judy, and Edouard Servan-Schreiber, "Circle of Friends: Social Network Analysis Helps Telcos Better Understand Customers," *Teradata Magazine*, March 2009, pp. 18–9.
6 Taneja, Nawal K., *Flying Ahead of the Airplane* (Aldershot, UK: Ashgate, 2008), pp. 20–7.
7 The Millennials are also referred to as Generation Y, Generation We, or even Echo Boomers as they are offspring of the Baby Boomers. The parameters of their age group differ among authors, whether it is those born between 1978 and 2000, 1980 and 2001, or just those born after 1980.
8 "Decoding the Digital Millennials," ResourceInteractive's *Litmus*, November 2006, p. 1.
9 "Europe Goes Nuts for Facebook," *CNN.com*, April 16, 2009.
10 Alsop, Ron, *The Trophy Kids Grow Up: how the millennial generation is shaking up the workplace* (San Francisco, CA: Jossey-Bass, 2008), p. 143.
11 ResourceInteractive, op. cit., p. 2.
12 Alsop, op. cit., p. 136.
13 "Managing the Facebookers," *The Economist*, January 3, 2009, p. 10 and "Generation Y goes to work," *The Economist*, January 3, 2009, pp. 47–8.
14 Fields, Bea, Wilder, Scott, Bunch, Jim, and Rob Newbold, *Millennial Leaders* (Buffalo Grove, IL: Writers of The Round Table Press, 2008), p. 210.
15 Alsop, op. cit., pp. 168–9.
16 Alsop, op. cit,, pp. 75–7, 87–8, 91–2.
17 Sheahan, Peter, *FL!P: How to Turn Everything You Know on its Head—and Succeed Beyond Your Wildest Imaginings* (New York, NY: HarperCollins, 2007), p. 154.
18 ResourceInteractive, op. cit., pp. 8–9.
19 Helm, Burt, "PNC Lures Gen Y With its 'Virtual Wallet' Account," *BusinessWeek.com*, November 26, 2008.
20 Ford Motor Company Press Release, March 26, 2009. (http://media.ford.com/article_print.cfm?article_id=30081)
21 The parameters of the Gen X age group differ among authors, whether it is those born between 1965 and 1976 or those born between 1965 and 1979.
22 Gravett, Linda, and Robin Throckmorton, *Bridging the Generation Gap* (Franklin Lakes, NJ: The Career Press, 2007), p. 44.
23 Gravett, op. cit., p. 41.
24 http://rismedia.com/2008-08-04/baby-boomers-the-forgotten-consumer-in-other-industries/

25 http://www.usnews.com/usnews/biztech/articles/050314/14boomer.htm
26 http://rismedia.com/2008-08-04/baby-boomers-the-forgotten-consumer-in-other-industries/
27 http://www.usnews.com/usnews/biztech/articles/050314/14boomer.htm
28 Gravett, op. cit., pp. 51 and 65.
29 In addition to those listed above, material for the Millennial chapter is derived from the following sources: Greenberg, Eric, with Karl Weber, *Generation We* (Emeryville, CA: Pachatusan, 2008); Grossman, David, "Are you a 'Next Generation' traveler?" *USA Today.com*; "On the Eve of the New Year," *Insight*, December 2008, Ypartnership; Tilin, Andrew, "Three Strategies for Managing Generation Y," BNET.
30 Charan, Ram, *Leadership in the Era of Economic Uncertainty* (New York, NY: McGraw-Hill, 2009), p. 16.
31 Quelch, John A., and Katherine E. Jocz, "How to Market in a Downturn," *Harvard Business Review*, April 2009, pp. 52–62.
32 Feldman, Amy, "Wooing the Worried: Companies are Reaching out to Skittish Consumers by Promising Breaks to Those Who Lose their Jobs," *BusinessWeek*, April 27, 2009, p. 24.
33 "From Buy, Buy to Bye-bye," *The Economist*, April 4, 2009, pp. 67–8.
34 McGregor, Jena, "There is No More Normal," *BusinessWeek*, March 23 & 30, 2009, p. 34.
35 "Trading Down: From Decadence to Discounts," *The Economist*, May 30th–June 5th, 2009, pp. 4–6 of "A Special Report on Business in America" and Flatters, Paul, and Michael Willmott, "Understanding the Post-Recession Consumer," *Harvard Business Review*, July–August 2009, pp. 106–112.
36 "On the Eve of the New Year," *Insight*, December 2008, Ypartnership.
37 Pilling, Mark, "Air New Zealand," *Airline Business*, August 2009, p. 49.
38 Kathryn B. Creedy, "Airlines Lag in Social Networking, Part I," in *Aviation Today's Daily Brief*, 13 April 2009.
39 Kathryn B. Creedy, "Airlines Lag in Social Networking, Part II," in *Aviation Today's Daily Brief*, 14 April 2009.
40 Kathryn B. Creedy, "Airlines Lag in Social Networking, Part III," in *Aviation Today's Daily Brief*, 15 April 2009.
41 Dunleavy, Hugh, WestJest Airlines, Presentation at the OSU International Airline Conference, Las Vegas, April 2009.
42 http://americanshelflife.wordpress.com/2009/04/05/virgina-america-vs-delta-airlines/

43 Slywotzky, Adrian, "Discovering and Creating Tomorrow's Customers" *Oliver Wyman Journal*, Spring 2009, p. 13.

44 "Airlines and Social Media: Carriers turn to Twitter, Facebook, Flicker, YouTube and online blogs," Centre for Asia Pacific Aviation, 27 August 2009.

45 Harteveldt, Henry, Forrester Research, Presentation at the OSU International Airline Conference, Las Vegas, April 2009.

46 Khosrowshahi, Dara, Expedia Inc., Presentation at the OSU International Airline Conference, Las Vegas, April 2009.

47 Tafoya, Shirley, Travelzoo, Presentation at the OSU International Airline Conference, Las Vegas, April 2009.

48 Taneja, Nawal K., *Flying Ahead of the Airplane* (Aldershot, UK: Ashgate, 2008), p. 110.

49 Pilling, Mark, "Air New Zealand," *Airline Business,* August 2009, p. 49.

50 Dougherty, Dave, and Ajay Murthy, "What Service Customers Really Want," *Harvard Business Review,* September 2009, p. 22.

51 Alamdari, Fariba, The Boeing Company, Discussion at the OSU International Airline Conference, Las Vegas, April 2009.

52 DiJulius, John R. III, *What's the Secret? To Providing a World-Class Customer Experience* (Hoboken, NJ: John Wiley, 2008), p. 50.

53 Griffin, Jill, *Taming the Search-and-Switch Customer: Earning Customer Loyalty in a Compulsion-to-Compare World* (San Francisco, CA: Jossey-Bass, 2009), p. 202.

54 Michelli, Joseph A., *The New Gold Standard: 5 Leadership Principles for Creating a Legendary Customer Experience Courtesy of The Ritz-Carlton Hotel Company* (NY: McGraw-Hill, 2008), pp. 171–2.

55 DiJulius, op. cit., p. 183.

56 DiJulius, op. cit., p. 183.

57 For an explanation of the distinction between customer service and customer experience, see following/next paragraph.

58 Schmitt, Bernd H., *Customer Experience Management: A revolutionary approach to connecting with your customers* (Hoboken, NJ: John Wiley, 2003), p. 161.

59 Selden, Larry, and Geoffrey Colvin, *Angel Customers & Demon Customers: Discover Which Is Which and Turbo-Charge Your Stock* (New York, NY: Penguin Group, 2003), p. 180.

60 Ballantyne, Tom, "Air NZ Takes on Recession Head-to-head," *Orient Aviation*, Vol. 16, No. 4, May 2009, p. 19.

61 Hyken, Shep, *The Cult of the Customer: Create an Amazing Customer Experience that Turns Satisfied Customers into Customer Evangelists* (Hoboken, NJ: John Wiley, 2009), pp. 102–3.

62 Griffin, Jill, *Taming the Search-and-Switch Customer: Earning Customer Loyalty in a Compulsion-to-Compare World* (San Francisco, CA: Jossey-Bass, 2009), pp. 192–3.

63 Schmitt, op. cit., p. 160.

64 Hyken, op. cit., p. 80.

65 Beemer, C. Britt, and Robert L. Shook, *The Customer Rules: The 14 Indispensable, Irrefutable, and Indisputable Qualities of the Greatest Service Companies in the World* (NY: McGraw-Hill, 2008) and Sharp, Isadore, *Four Seasons: The Story of a Business Philosophy* (NY: Portfolio, 2009).

66 Mount, Ian, "JetBlue Founder's Revenge: A New Airline," *CNNMoney.com*, March 20, 2009.

67 DiJulius, op. cit., p. 36.

68 Heskett, James L., Jones, Thomas O., Loveman, Gary W., Sasser, W. Earl Jr., and Leonard A. Schlesinger, "Putting the Service-Profit Chain to Work," *Harvard Business Review*, July–August 2008, pp. 126–7.

69 Ibid., pp. 118–129.

70 DiJulius, op. cit., p. 119.

71 DiJulius, op. cit., pp. 176–7.

72 DiJulius, op. cit., pp. 178–9.

73 Neuborne, Ellen, "The Latest on Loyalty," *ANA Advertiser*, April 2009, pp. 40–43.

74 Hyken, op. cit., pp. 24–6.

75 Michelli, op. cit., pp. 89–90.

76 Griffin, op. cit., p. 200.

77 "American Airlines Pays Employees $200 Bonuses," Centre for Asia Pacific Aviation, Date posted: 18-May-09.

78 Hyde, Paul, and Barbara Bacci Mirque, "Jim Garrity: Gold-Standard Customer Experience," in *CMO Thought Leaders: The Rise of the Strategic Marketer* (Edited by Geoffrey Precourt and Published by Booz Allen Hamilton, 2007), p. 143.

79 Perman, Stacy, *In-N-Out Burger: A Behind-the-Counter Look at the Fast-Food Chain that Breaks All the Rules* (NY: Collins Business, 2009).

80 Hawn, Carleen, "The In-N-Out Burger," *Fast Company.com*, December 19, 2007.

81 Tisch, Jonathan M., with Karl Weber, *Chocolates on the Pillow Aren't Enough: Reinventing the Customer Experience* (Hoboken, NJ: John Wiley, 2007), p. 28.

82 In-N-Out Burger is a privately held corporation; precise financials are not available.

83 DiJulius, op. cit., p. 153.

84 Zappos corporate website: www.zappos.com (http://about.zappos.com/meet-our-monkeys/tony-hsieh-ceo)

85 McGregor, Jena, "When Service Means Survival," *BusinessWeek*, March 2, 2009, pp. 29, 32–3.

86 Presentation by Tony Hsieh, CEO, Zappos at ANA Conference, October 17, 2008.

87 Hyken, op. cit., pp. 70–1.

88 Presentation by Tony Hsieh, CEO, Zappos at ANA Conference, October 17, 2008.

89 DiJulius, op. cit., p. 55.

90 Zappos corporate website: www.zappos.com (http://about.zappos.com/our-unique-culture/zappos-core-values/create-fun-and-little-weirdness)

91 Presentation by Tony Hsieh, CEO, Zappos at ANA Conference, October 17, 2008.

92 McGregor, Jena, "Zappos' Secret: It's an Open Book," *BusinessWeek*, March 23 & 30, 2009, p. 62 and Hyken, Shep, *The Cult of the Customer: Create an Amazing Customer Experience that Turns Satisfied Customers into Customer Evangelists* (Hoboken, NJ: John Wiley, 2009), pp. 68–9.

93 Zappos corporate website: www.zappos.com (http://zeta.zappos.com/ViewProduct.action?productId=7496010&colorId=1)

94 In addition to those listed above, material for the Zappos case study is derived from the following sources: Coster, Helen, "A Step Ahead," *Forbes.com*, June 2, 2008; McFarland, Keith, "Why Zappos Offers New Hires $2,000 to Quit," *BusinessWeek.com*, September 16, 2008; O'Brien, Jeffrey, "Zappos Knows How to Kick it," *Fortune*, February 2, 2009, pp. 54–60; Taylor, Bill, "Why Zappos Pays New Employees to Quit-And You Should Too," *HarvardBusiness.org* (http://blogs.harvardbusiness.org/taylor/2008/05/why_zappos_pays_new_employees.html), May 19, 2008.

95 Online edition of *The New York Times*, 23 July 2009.

96 Stoller, Gary, "Hotel Service Improves Despite Cost Cuts," *USA Today*, July 29, 2009, p. 3B.

Chapter 4
Innovating around Airline Realities

Innovation involves looking at new ways of doing things. Although innovations can be brilliant inventions, they often emerge from just looking at a situation from an alternate perspective. Specifically, innovation can take on many forms and can occur at many levels, such as direct changes to a product or service. Examples include 3M with its sticky notes, Qantas with the Business Class in long-haul markets, Virgin Atlantic with the in-flight entertainment at individual seats, Hitachi with the Plasma HDTV, Lufthansa with the all Business Class transatlantic flights, and Amazon with the Kindle e-reader (a device to digitally read books, newspapers, and magazines). Or an innovation can change an entire industry, just as Apple's iPod changed the music industry, Cirque du Soleil, the Canadian live entertainment company, changed the circus industry, and Nintendo's Wii changed the video game industry. Yet another form of innovation involves revamping the way in which business is traditionally conducted within an industry. Consider how Enterprise Rent-A-Car changed the car rental landscape by segmenting the market and targeting those in need of transportation due to car repairs as well as travelers, and by offering alternate distribution channels (storefronts in the suburbs rather than kiosks at the airports). Moreover, how about Zipcar, the car-sharing startup, that has totally elevated the concept of personal mobility. Think about how eBay, the online conduit for buying and selling goods, literally turned upside down the traditional second-hand sales channels, such as flea markets and garage sales. EBay's innovation allows one to buy or sell almost anything, anytime, and anywhere in the world.

Innovation in the airline industry is more complex than in many other industries due to the existence of numerous constraints such as strict government regulations (relating not just to safety, but also to business operations), limitations of the infrastructure, and government mandated and operated facilities at airports dealing with immigration and, more recently, security. Yet, despite these constraints, some airlines have managed to adopt new business models. One example, Ryanair, changed the traditional business model by not only serving major markets through the use of smaller and less conventional airports, but also by charging less for the basic transportation and by creating profit through the generation of ancillary revenue.

This chapter will provide examples of companies implementing innovations at a variety of levels, all resulting in increasing margins with alternative value propositions that are different from those offered by competitors. This chapter begins with some insights into the major drivers behind innovation, then moves on to business model innovation, and concludes with several case studies of companies who have excelled in the area of innovation. Each section provides some associated insights for airlines.

Major Drivers of Innovation

In researching the best practices and competencies relating to innovation, there appear to be four major drivers behind the successful implementation of innovation.[1] See Figure 4.1.

Challenging Conventional Thinking

The first driver, challenging conventional thinking, relates to producing an environment where employees feel they are able to explore freely. However, as Judy Estrin notes, the manner in which management utilizes the questioning process (for example, questioning assumptions) can affect behavior and motivation among employees as well as set the tone for the whole organization. For example, does management question in an inquisitive manner or in a judgmental manner? The questioning approach can alter

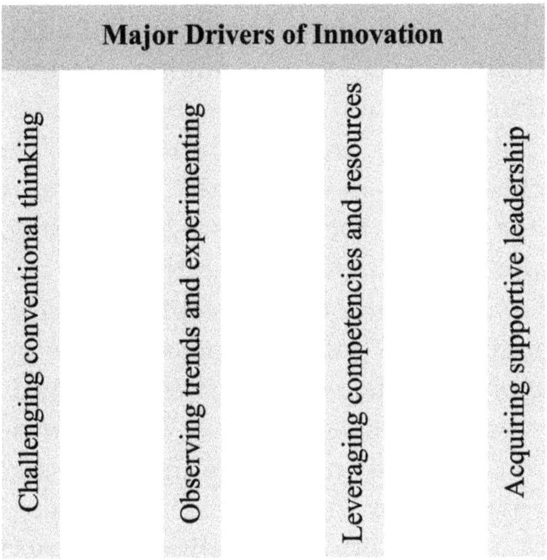

Major Drivers of Innovation

Challenging conventional thinking

Observing trends and experimenting

Leveraging competencies and resources

Acquiring supportive leadership

Figure 4.1 Major Drivers of Innovation

the attitude of employees. FedEx is an example of a company that recognizes the importance of the sense of curiosity on the part of its employees, whether on the job or at home. This innate sense of curiosity inspires innovating employees to dig deeper and uncover potential that others have overlooked.

Challenging conventional thinking through questioning, or, as some business writers such as Skarzynski and Gibson call it, challenging orthodoxies, can drive success within a company or an industry. IKEA challenged orthodoxies in the furniture industry by offering standardized products rather than custom-made products, and implementing a "take and carry" system rather than offering the traditional, assembled furniture that was delivered to a customer's home. Nintendo challenged conventional thinking in the video game industry, as discussed in the case study later in this chapter, by moving away from competing on the traditional factors of graphics quality and game-play intensity, as well as by making their Wii offering more physically interactive and socially engaging for a variety of users. While orthodoxies are not necessarily bad or wrong, problems can arise when their relevance has passed and/or they hinder, rather than foster, progress.

With that being said, how does one challenge conventional thinking or orthodoxies? One way might be to go in an opposite direction of others in the industry. An example of a company that pursued this path is the grocer, Whole Foods. Specifically, the organization chose to go in exactly the opposite direction of the traditional, large, chain grocery stores by focusing on a strategy of nutrition and authenticity rather than on convenience and price. Whole Foods actually charges a premium for fresh, natural food that is obtained, as much as possible, from local sources. By challenging orthodoxies, new marketplaces and new offerings can be discovered, complete with new rules and new structures. Skarzynski and Gibson acknowledge that even small details can present opportunities for innovation. Specifically, they provide an example from within the hospitality industry. Why is it that even if a guest checks into a hotel very late at night, the guest still must check out by noon the next day? Why can hotels not operate like car rental agencies in terms of paying for a room for 24 hours as one does when renting a car, rather than have specific check-out times? Admittedly, such a change would require numerous operational changes, such as with respect to housekeeping services and their associated costs.

Regardless of how successful an organization is, it cannot afford to be complacent and thinking it knows all. The idea is to raise questions and keep the staff open to new insights about consumer behavior to gain a true understanding of how consumers interact both physically and emotionally with their environment as well as products and services within that environment. While every company claims to listen to its customers, it is the type of feedback gained from observing customers in a real setting that improves customer satisfaction through innovations. An example given by Tom Kelley of IDEO is that of traveling through Charles de Gaulle Airport to catch the train to connect to the metro subway system. He notes that while the train itself is fine, the experience at the turnstiles to enter the train leaves something to be desired. Specifically, the turnstiles' configuration does not take into consideration that those arriving from international flights will have luggage that they will need to maneuver while trying to pass through the turnstiles in order to board the train. While this may not be a huge issue in the grand scheme of things, he notes that there is certainly a better way to handle the situation. Specifically, if the time had been taken to watch people try to get through the turnstiles with all their bags or to anticipate their needs, this simple, but frustrating, situation could have been avoided.

One useful tool to encourage and facilitate innovation is to cut across different corporate functions and bring diverse groups together to achieve multidisciplinary solutions. Again, the grocer, Whole Foods, provides an example. The company's workforce is organized into eight teams per store. Each team is responsible for its own hiring and employee retention, and has input in terms of what is stocked, how it is displayed, etcetera. The teams also meet regularly to collaborate on issues and to solve problems. It has been said that collaboration is the lifeblood of Whole Foods' operations. In cutting through the cross-functional barriers, the opportunity is created to build upon and enhance different individuals' talents to drive the innovation process. Consider Kelley's example of an executive at the toy maker Mattel, who organized a team which launched a new process that, in turn, created a US$100 million girls' toy platform in just three months. Specifically, the executive obtained a special space for the project on the company's campus, then assembled a cross-sectional team from the current Mattel employee pool, thus pulling them from their current roles for twelve weeks. The group engaged in a variety of activities in an effort to gather insights that ultimately led to the innovation of the new toy platform.

Innovative companies cross barriers and overcome independent silos in an effort to encourage collaboration and spur innovation. By breaking down barriers within organizations as well as exploring how other industries conduct business, innovators have the unique ability to bring minds together that may never have otherwise had an opportunity to collaborate, and implement innovations that, at first, may not have seemed to have anything to do with the task at hand. For example, a hospital implemented a new logistics system in terms of the manner in which it transported patients based upon insights gained on a visit to a taxicab dispatch office. Finally, innovative companies also study other industries and cultures, and then apply those findings to their organization. Procter & Gamble is an example of a company that has had success with cross-functional and multi-industry insights, as noted by Kelley. Specifically, P&G's laundry products business took its knowledge of safe, whitening agents and collaborated with its oral hygiene business, yielding the innovation of the home teeth whitening system, Crest Whitestrips. Another example, also at P&G, involves its PUR division combining its knowledge of water purification with its expertise knowledge in spot removal from its Cascade dishwasher detergent to create a carwash system, Mr. Clean AutoDry.

In the process of questioning and challenging conventional thinking, a lot of learning comes from trial and error, and failure is part of the process of innovation. An example given by Kelley is BMW. Specifically, a few years ago, BMW decided to forgo its traditional advertising channels (such as thirty second commercials), and instead created and implemented eight minute short films to be viewed online at bmw.com. While BMW was unsure whether the experiment would succeed, the organization went ahead, and the innovative short films turned out to be a huge success. Moreover, the experiment yielded great marketing buzz. Specifically, not only did BMW gain media coverage around the world, but car enthusiasts forwarded the link to all of their friends, generating enormous traffic to BMW's website. The films have even surfaced on in-flight entertainment systems.

Airline Examples Relating to the First Pillar of Innovation

The following are some examples of airlines whose managements challenged orthodoxies and decided to operate with different models.

- Ryanair questioned the wisdom of serving conventional airports and having standard agreements with airports. Instead, it based its operating model on the use of unconventional airports and negotiated a subsidy provided by airports, related businesses, and local communities for bringing traffic to their communities.

- Qantas questioned the assumption that an airline had to set up a hub-and-spoke system, not only centrally located geographically, but even more important that it be in the home country of the airline. Given the geographic location of the population in Australia (mostly at the edges of the country), a conventional hub-and-spoke did not make much sense. Qantas used a different business model to develop a hub-and-spoke system off-shore, one based in Singapore.

- Allegiant Air questioned whether it was necessary to provide daily service and high frequency in US domestic markets to have viable and profitable airline operations. Instead, the airline developed its "scheduled charter" services model based on the one in Europe where the charter airlines provided service to beaches that had become "industrialized."

- AirAsia questioned the premise that it was not feasible to have low cost, low fare service within markets in Asia. It also questioned why it was not possible to have operating costs even lower than Ryanair, probably the lowest cost operator of scheduled services in the world. It based its model on Ryanair, but drilled down even further into various components of costs, and managed to beat even Ryanair.

- Continental Airlines took on the assumption that narrow-body aircraft could not compete effectively with wide-body aircraft in long-haul intercontinental markets. It scheduled Boeing 757 aircraft in transatlantic service and succeeded sufficiently to encourage competitors to take similar initiatives.

- Flybe Airlines, a large regional carrier based in the UK, questioned the assumption that turboprop aircraft could not compete effectively against standard jet aircraft. It has been successful in flying the advanced turboprop aircraft (Bombardier's Q400s) within Europe in competition with conventional jet aircraft.

Observing Trends and Experimenting

The second driver involves a very detailed analysis of trends that can dramatically change the competitive rules or industry structures and lead to new opportunities. This is an area where business intelligence needs to be taken to a higher level, as discussed further in Chapter 5. Observations must be based on real experiences (not based on reports or focus groups), as Skarzynski and Gibson illustrate in their examples of both Nokia and Best Buy. In the early 1990s, Nokia began to take note of the emergence of a global youth culture. However, rather than just reading about this emerging group in trend reports, Nokia decided to actually send a group of engineers out to observe the group firsthand. The group of engineers observed the youth culture at play, such as at Venice Beach in California, King's Road in London, and in the club scene in Tokyo's Roppongi district. In doing so, the engineers were able to collect valuable insights that they could take back to Nokia, thus putting the company at the cutting edge of the industry.

Consumer electronics retailer, Best Buy sought out some unique places in other parts of the world that they had never looked before to gain some insights about the future; one of those places being South Korea. On nearly every corner in South Korea, one can find a *Baang,* which is a venue where one can play video games, a popular recreational activity in the country. In visiting these venues and observing the patrons, Best Buy was able to obtain ideas for innovative new strategies, aimed at consumers who enjoy keeping up with the cutting edge of electronics and gaming. Such strategies are imperative for Best Buy to stay ahead of its competition.

Observing trends and experimenting can help companies understand unarticulated consumer needs and ways to satisfy an unmet need or a current frustration of a consumer. Focus groups and other traditional market research methods are not usually successful tools to ascertain such information as the participants are not necessarily aware that they have these needs, yet. For example, no one was necessarily requesting a comfortable "third place" outside of home and work to relax or socialize with friends over a specialty beverage, yet Starbucks filled a need for such a place in peoples' lives. Similarly, Amazon.com provided an online store where one could find and purchase nearly any book ever published, complete with reviews by other fellow readers; a totally new concept to consumers. These companies were both highly successful in finding the space where there were unmet customers' needs that they could address. Similarly, Cirque du Soleil and Nintendo, companies highlighted in the case studies at the end of this chapter, also have been successful in finding such open space and in addressing such unmet needs. It is important to note that needs may be both tangible, such as mobility, and intangible, such as being associated with an organization whose products are environmentally friendly. This is one problem with traditional market research; consumers can only tell you what they know, not what they do not know. Admittedly, it is difficult to analyze markets that do not currently exist from the viewpoint of market size and financial returns as required by investment processes. In the case of airlines, it is not only the use of resources, but also the creative ways to get around the inherent constraints that impact their business practices.

Furthermore, in addition to anticipating and attending to customer needs, innovators also need to ensure that a customer is treated as a person, rather than a number, when conducting business with the organization. For example, as pointed out by Kelley, Bank of America formed a new group within its organization, the Innovation & Development team, which selected 20 branches of the organization to use as test sites to implement innovations targeted to anticipate the needs of customers. Such innovations included the addition of comfortable couches to give customers a place conducive to reading about investment products and financial magazines, an "investment bar" with computers to allow customers the ability to check their accounts or access the Internet, and electronic stock tickers and flat-panel television sets tuned to CNN to provide entertainment and information to customers while waiting in line. The goal was to create more of the intimate feeling of boutique, private bank rather than the branch office of a large, faceless organization. The personalization aspect can also be conveyed in the form of stories. Specifically, a story can be used to build both internal morale, as well as external awareness. Kelley provides the example of

Medtronic, a medical technology firm, which utilizes stories told by the patients themselves to demonstrate how the organizations' products have changed, and in some cases, saved lives.

Finally, companies must implement empathy, the ability to reach outside of oneself and connect with others, in their quest for innovation. Author Dev Patnaik recognizes that companies find success when they put aside their own problems and start caring about everyday people; companies need to walk in their customers' shoes. The author gives the example of Nike, who has literally done that, walk in its customers' shoes. Specifically, he describes the work of an employee at Nike, a man who literally spends his time going to high schools all over in order to attend gym classes, track meets, and races to meet with athletes. "Why?" one might ask. Because he recognizes that it is imperative that Nike understands what this consumer group is truly looking for in a running shoe. He even shops where high schoolers shop, and hangs out where they do. The idea is to truly walk in their shoes so that he may think and feel the way they do and make sure that the company is catering to their needs with their product offerings. When Nike felt that it was slipping in the college marketplace, he actually went to college campuses to observe what the students were doing, and what their current needs were in an athletic shoe so that the company could adjust accordingly.[2] How many companies engage in such activities and go to such lengths to learn about their consumers needs?

Airline Examples Relating to the Second Pillar of Innovation

Following are some airline examples relating to the "observing trends and experimenting" driver.

- Air Canada unbundled its product relating to its pricing structure with an "à la carte" pricing system that offered transparent, price-service options. This strategy in some ways reversed the commoditization of commercial air travel. For more on the concepts of price bundling and unbundling, see Chapter 5.

- Lufthansa introduced (a) all Business Class aircraft across the Atlantic, (b) private aircraft to and from smaller cities to connect with Lufthansa's intercontinental flights for passengers traveling in Premium Class, and (c) totally separate check-in facilities for Premium Class travelers in intercontinental markets.

- All Nippon made it possible for its domestic passengers to make reservations and do ticketing and check-in, all via a computer or a mobile phone (with the boarding passes at the self-check machines at the airport).

- Southwest experimented with service to major, conventional airports, and refined its strategy to compete effectively with major, incumbent carriers.

- British Airways decided to experiment with a small all Business Class aircraft between London City Airport and New York. The limitation on the fuel that could be carried on board the proposed Airbus 318 (with 32 flat-bed seats) requires a fuel stop in Shannon, Ireland. However, the time on the ground in Shannon can be used to pre-clear US immigration and customs, enabling passengers to arrive at a JFK terminal as domestic travelers.

Leveraging Competencies and Resources

The third driver, leveraging competencies and resources, involves just that, leveraging what a company knows how to do well along with resources it can bring to bear; brands, patents, infrastructure, customer databases, etcetera. For example, everyone thinks of Google as a search engine, which it certainly is; however, the company's underlying software code is also a powerful strategic resource that can be leveraged in a variety of ways, such as targeted advertising. The key insight from Google is that it constantly builds on its competencies and uses its resources to not only sustain its existing technologies, but also to invest in new (even disruptive) technologies so as to not lose its market dominance. As discussed later in this chapter in the case study on Google, the insight is not for companies to continue to use their competencies and resources just to provide customers with products and services that they need in order for the company to stay in business. It is also to invest in ideas for new products and services that customers may not even know that they want, as addressed in the discussion on driver number 2.

An important part of the third driver, allocation of resources, is to get wide employee involvement. While many companies indicate that they would like everyone in their organization to be involved in the innovation process, Whirlpool truly walks the talk. Given that the company has a global workforce of seventy-three thousand people, this could be quite a challenge, but the company was careful in its consideration and implementation of a program that was appropriate for its vision of innovation from everyone, everywhere in order to make it a reality. The company was careful in choosing a program to bring their vision to life, remembering that it was imperative to innovate within their brand focus or risk eroding or even destroying that advantage. Specifically, in the first year of Whirlpool's innovation initiative, the company implemented a program that involved three teams of twenty-five people each (from around the world) located in three, regional headquarters: São Paulo, Brazil; Comerio, Italy; and Benton Harbor, Michigan. These people were the first groups to work with Strategos, an organization that specializes in teaching a set of concepts, tools, and methods that foster innovation within a company. These seventy-five employees actually temporarily left their traditional jobs within the company to move to one of the three regional headquarters to become innovators. The original idea was that they learn the innovation concepts, tools, and methodology, and then roll it out to the rest of the organization in a specific manner. One third of the group would return to their jobs and implement the innovation tools in their daily work. One third were to run business units that were launched based on innovation results, and a third were to teach the tools and skills to other employees.

The initial program was a success; there became a great demand from employees and teams who wanted to learn innovation techniques, and twenty-five people were not enough to fill this demand. The company wanted to school as many individuals as possible in terms of innovation techniques to be able to implement innovation themselves, as well as mentor others in the innovation process. To accomplish this, Whirlpool, along with Strategos, created a skill set for these individuals, who are referred to as "I-mentors." The "I-mentor" program includes a rigorous training program outlining the company's innovation process, tools, and techniques. Each candidate is required to work on several innovation projects in order to be able to apply the tools in a "real" setting before she can apply to become a certified "I-mentor." The certification process is also rigorous, and involves peers evaluating the skills, application, and results of the "I-mentor." The training program includes, but is not limited to, the following elements. First, the program links innovation to other Whirlpool processes, as the company recognizes that it is critical for its employees to understand how to align the innovation process with other systems such as finance, product development, and manufacturing. Second, it is instilled that customer insights, as well as trends in customer behavior, are a good basis for creative ideas and innovation. The company recognizes that mentors must learn and know this as, in the past, the company tended to focus too much on features and engineering design which did not align with customers in terms of their needs, wants, insights, or trends. The mentors are also taught voice-of-the-customer tools, as well as how to translate ideas into customer requirements in terms of products, services, features, and even the buying experience. Third, the mentors are trained to create a business plan for innovation. Specifically, this training not only includes a plan for experimentation, but also redefines the business model based on actual experience in the market.

Whirlpool now has more than eleven hundred people who have been trained as "I-mentors." These "I-mentors" come from a variety of functional backgrounds, including accounting, engineering, finance, human resources, manufacturing, and technology. This hybrid of positions helps to foster the company's vision of innovation from everyone, everywhere.[3]

Airline Examples Relating to the Third Pillar of Innovation

The following are just two examples of airlines that have leveraged their competencies and resources. One example of leveraging competencies is the intercontinental network of TAP Portugal that focuses on the strengths and profitability of the airline more than on high visibility and glamour. As seen in Figure 4.2, TAP Portugal has focused its intercontinental network on one of its resources; its colonial ties. Instead of flying all over the world (including countries that are in vogue, such as India and China), TAP offers nonstop flights from Lisbon, Portugal to eight destinations in Brazil (see the route map in Figure 4.2). Typically, European airlines serve only the two major destinations in Brazil, São Paulo and Rio de Janeiro. Complementing the extensive service to Brazil, TAP focuses on six destinations in Africa. It is interesting to note that while the general perception of Africa is that it is a continent with extreme poverty, TAP manages to have a viable operation flying to relatively poor countries such as Mozambique (with GDP per capita estimated to be just under US$1,000). Flying to countries in Africa also circumnavigates the regulatory constraint in which governments, such as China, control airline market entry and frequency (for more on Africa, see Chapter 6).

Similarly, Turkish Airlines is leveraging its location and the size of its country to expand its international operations. For some connecting markets, the location of Istanbul as a connecting hub is better than those located in the Gulf region of the Middle East. Moreover, the size of Istanbul is much bigger (with population approaching 14 million) than the size of the new hub cities in the Gulf region of the Middle East. Chapter 6 provides more information.

Acquiring Supportive Leadership

The fourth driver of innovation, supportive leadership, has five elements. First, there is a need for the leadership to create an environment in which employees are given assurance that there will not be repercussions for taking risks that may lead to failure, as long as they have been diligent in understanding why their

Figure 4.2 Intercontinental Route Map of TAP Portugal
Source: Shown in the Airline's Website.

initiatives did not succeed. As mentioned earlier, failure is an inherent part of innovation and innovating companies tolerate it at some level as an investment eventually leading to a winning outcome. For example, Estrin points out that while Steve Jobs of Apple and Fred Smith of FedEx have extremely different leadership styles and have achieved success in completely different industries, they do have one common characteristic; they are both very willing to take bold, yet calculated, risks. In the process, both have endured open and expensive failures—Apple's Lisa computer and FedEx's ZapMail service. However, both still are willing to take on risk and continue to eagerly pursue new ideas.

The second element of leadership is that it needs to promote openness, an atmosphere that encourages and inspires employees to think out of the box, to imagine, and also to collaborate with one another. Again, Estrin points out that by sharing information early on during a project, it allows for valuable feedback, which can, in turn, catch issues early on in the process. For example, she indicates that some of the most costly failures in Silicon Valley were the result of projects that had been kept behind closed doors. Openness also relates to the physical environment. Kelley gives the example of Levi Strauss & Co., which was once located in a high-rise building. While there was nothing wrong with the office itself, the space just did not fit the needs of the company. Specifically, management felt that the space did not support and promote the organization's family atmosphere. Thus, management secured a new environment for the organization's headquarters—a campus-like setting. This campus setting which included low-rise buildings, outdoor terraces, and a park, promoted an openness to foster the family atmosphere and collaborative nature of the organization. Google has also considered the physical environment in terms of fostering innovation, as described in the case study on Google later in this chapter.

The third element of leadership, closely related to openness, is patience. It calls for an organization with employees who are comfortable with dealing with ambiguity to foster the full innovative process, rather than jumping on the first idea or solution that comes along. An example made by Estrin involves two companies, 3Com and Cisco, competing for a piece of the networking market in the late 1990s. During this time, 3Com's board decided to pull out of the business as they wanted a quicker means of achieving profitability. It has been estimated that this lack of patience has set the company back years, and while few people remember the name 3Com, Cisco has become a prominent name in the networking industry.

Innovation takes time. It can take numerous cycles of brainstorming, testing, analyzing, and adapting in order to get a project just right. For example, Estrin notes that Netflix was originally launched by a frustrated individual who was tired of paying late fees when renting movies. The original offering required users to pay US$4 to rent a single DVD for a week, plus a US$2 shipping and handling fee. Soon, the founder realized that the offering was not a "wow" for consumers, and, therefore, he went back to the drawing board to brainstorm a "wow" breakthrough. In further analyzing the movie rental business, he concluded that a monthly subscription model, complete with the implementation of a "dynamic queue" system (customers could always have a DVD to watch and more lined up in queue) could be the answer. Netflix re-launched in 1999 with its new innovation, which was definitely successful at achieving the desired "wow" in the marketplace.

The fourth element of leadership relates to trust, where the leadership of a company believes in both its people, as well as the innovation process itself. It also requires that the employees believe in themselves in order to possess the ability to plow through obstacles and keep the vision, but to also remain open to new ideas and questions along the way. Estrin gives the example of a Best Buy employee, who refers to this as living in both the worlds of self-confidence as well as self-doubt. The vision component is also illustrated by Jeff Bezos, the CEO of Amazon, who is determined to reinvent the book business. He accomplished this by first creating a new way to sell books and now through a new way for people to read books and newspapers, namely, the Kindle e-reader.[4]

The fifth element of leadership is about the balance of responsibilities at the top. While most business analysts agree that innovative companies tend to have strong leaders, some business analysts have gone deeper into the characteristics of successful leadership. Let us consider an industry that relies very heavily on innovation—the fashion industry. Why is this? First, because fashion companies are, in essence, reinventing themselves each season by offering a new "must have" product line that ignites demand for the new, but also makes the old obsolete. How are fashion companies able to do this season after season, year after year? That question leads to the point about leadership which, according to Rigby, Gruver, and Allen, is that at the top of nearly every fashion company, there exists a distinctive type of partnership in which one partner is the creative director and the other partner is the business director. Specifically, the creative director is a right-brain individual who is imaginative and possesses the ability to continuously churn out new ideas that cater to the wants and needs of the company's target audience. The business director, on the other hand, is a left-brain individual who possesses the ability to make the hard decisions based on thorough analysis. This form of partnership can be found at famous fashion houses, such as Chanel and Gucci. However, these both-brain partnerships have been found outside of the fashion industry, as well. Some examples include Apple, BMW, Hewlett-Packard, and Procter & Gamble.[5]

The bottom line is that leadership must not be distracted by clutter and noise that take up time and energy, leaving little or nothing to pursue innovation. Additionally, the leadership must not allow the organization to become so used to the routine and systematic way of doing things that they are unable to see the benefits possible through even slight deviations, in terms of innovation. Within traditional workplaces, the silo effect can truly become a limitation for innovation, just as do fixed roles and rigid disciplines. This is true as much of innovation occurs at the intersection of the boundaries between disciplines, departments, and roles. Leadership, therefore, is not just the fourth driver, but also the enabler of the first three drivers of innovation. It is leadership that sets a tone within the organization that fosters a culture that embraces innovation as exemplified by the experience at Apple, Nokia, Procter & Gamble, Toyota, and Virgin Atlantic. In the case of the latter, it was the leadership of Sir Richard Branson that supported continuous innovation, starting with in-flight entertainment systems in Economy Class cabins, customer experience in the airport lounges, and Premium Class cabins in intercontinental flights. Management knew that these innovations would be copied quickly, but the key difference is that Virgin's innovative culture kept Virgin Atlantic one step ahead of competition, and gave the airline, itself, the label of being innovative.

An Airline Example Relating to the Fourth Pillar of Innovation

Sir Richard Branson is one clear example of supportive leadership within the airline industry relating to innovation. Virgin is a reputational brand, based on the reputation of Sir Richard Branson and, in turn, on his focus on the customer in all of the businesses within the Virgin Enterprises, from Virgin Atlantic to Virgin Brides. Sir Richard believes that customers need more than just value, but also choice, particularly where choice does not exist. He believes in customer service and customer experience, and that the key driver behind these two related concepts is innovation. Innovations, in turn, rely on superior customer understanding, and that, in turn, comes primarily from employees. In the case of Virgin Atlantic, it was management's support of innovation that led to the name Upper Class given to what is basically a First Class product at Business Class fares. It was the first airline to offer limousine service, the first to offer check-in service in the airport's parking area, the first to offer in-flight entertainment to individual passengers at their seats, and the first to offer unconventional airport passenger lounges.

As examples of managerial support for experimenting, think about the following to initiatives. First, in addition to the limousine service to and from the London Heathrow Airport, Virgin tried the use of motor bikes, and even boats on the River Thames in an attempt to overcome congestion on the roads. Second, Virgin offered tailor-made suits for passengers traveling between London and Hong Kong. Measurements would be taken prior to boarding in London and sent to tailors in Hong Kong while the passenger was en-route, then the suit would be delivered to the passenger upon arrival in Hong Kong.[6] Think of the word-of-mouth buzz that such a marketing initiative generated. All of these innovations stem from employees who were supported by Sir Richard to think openly about the needs of their customers. The key is that top management is aware of the internal and external customers. In order to achieve innovation, management must keep the employees engaged and ensure that employees are treated as potential sources of ideas.

Business Model Innovation

The key aspect of business model innovation is to create value, both for the customer (providing a solution to a customer's need) and the company (making a profit). In the case of an airline, consideration must also be given to the existence of industry-related constraints. So, how does a company build a great business model that adds value for customers and itself? In their article in the *Harvard Business Review*, Johnson, Christensen, and Kagermann illustrate how two companies, the Tata Group and Dow Corning, derived game-changing business model innovations.[7]

Specifically, as stated in Chapter 2, Ratan Tata of the Tata Group noticed how whole families, including children, were traveling on motor scooters in Mumbai in the rain, and saw a need, the need to offer a safer means of transportation for these families. The idea was to provide transportation that would shelter them from the elements, but would also compete with the price point of a scooter.

After determining this customer value proposition, Tata addressed the profit formula component. The price factor was a huge barrier in this situation, as the scooter owners were not car owners due to the fact that the least expensive car available in the country at that time was approximately five times the price of a scooter. Tata went completely the other way and considered what it would take to offer a car that was approximately half of the price of the least expensive car currently being offered in India. In order to make this happen, a drastic reduction in both the cost structure and in the gross margins would have to occur. Tata believed that the potential consumer segment was significant, and, therefore, the company could still turn a profit through a potentially great sales volume. Next, Tata had to identify the key resources and processes. In order to achieve its customer value proposition as well as its profit formula for the proposed Nano vehicle offering at US$2500, the entire means of design, manufacturing, and distribution would have to be changed. One way that Tata achieved this is by utilizing a group of young engineers who would not be burdened by the company's existing ways, especially the existing profit formulas, as the more experienced engineers might have a tendency to do. This group was able to achieve the task at hand by implementing such innovations as minimizing the number of parts in the vehicle, using fewer vendors, and outsourcing a large percentage of the Nano's components, all yielding less transaction costs yet better economies of scale (for more on Tata, see Chapter 2).

The second example provided by Johnson, Christensen, and Kagermann is the global silicones supplier, Dow Corning. Dow Corning had been in the business of selling silicone-based products and providing highly technical services to a variety of industries for many years. However, studies indicated that customer needs were changing. Specifically, many customers had become more experienced in terms of the silicone application, and therefore those customers no longer required the technical service aspect of Dow's offerings. Instead, customers now had the need of basic products offered at low prices. Dow had to determine a way to be able to serve this new need.

A team was formed to determine how to proceed with this new business need. The team first considered the customer value proposition. Specifically, it was determined that the price point had to decrease by 15 percent in order to meet the needs of these consumers. The team determined that to reach this new price point, a new profit formula would be required; one that included a lower cost structure. This, in turn, required the development and implementation of a new IT system so that the product could be offered at bulk prices through the internet (unlike the high-end offering that involved customized solutions and negotiated contracts). The key resources that Dow had in place were highly trained in delivering a customized experience, given the mature and successful state of the company. The new business would require different resources, ones much more standardized as well as automated. The leader of the team had to determine whether this new business venture could be successful with Dow's existing infrastructure. Therefore, he implemented a test "war game" to determine how existing employees and systems would handle the new customer value proposition. Old habits and procedures did not allow for the needed change. Therefore, a new business unit was implemented so that the new consumer need would not be burdened by the company's existing ways. The traditionally high-margin company gained a new opportunity in terms of a low-margin offering, and established separate business units to meet the drastically different needs of these offerings. As Johnson, Christensen, and Kagermann note, established companies are not usually successful in terms of implementing dramatically new offerings unless the company takes the time and effort to truly understand how the new opportunity relates to the current business model, which is exactly what Dow did and, therefore, why it was successful.

What is the degree and nature of change required for business model innovation? Johnson, Christensen, and Kagermann illustrate five circumstances that signal a need for a change in an organization's business model.[8]

- There is an opportunity to extend an offering to a large potential segment that currently is not able to be consumers in the marketplace due to expense. An example would be Tata's low cost Nano car capturing part of the motor scooter segment (for more on Tata, see Chapter 2).

- There is an opportunity to hone in on cutting-edge technology by wrapping a new business model around it. An example would be Apple's phenomenal business model that provided a winning combination of both hardware and software, as well as service.

- There is an opportunity to focus on a job-to-be-done rather than focusing on products or services, or the features related to the product or service offerings. An example would be FedEx, who entered the package delivery market on the premise of getting the job done much faster and more reliably, rather than on trying to compete by offering lower prices on its service.

- There is a need to fight off low end disrupters in the marketplace. An example would be if Tata's low cost Nano automobile is a success, its presence could be a threat to other automakers.

- A business model change might be required in response to a shifting basis of competition. For example, high end construction power tool manufacturer Hilti (based in Liechtenstein) needed to implement a change to its business model as low end entrants began to offer tools that were "good enough," and thus started to eat into Hilti's market share.

Decades ago, some businesses developed a "side door" approach to make money. This approach involved the sale of complementary products. Gillette was a pioneer in this concept by offering one product, the razor, relatively inexpensively, but then offering a necessary complementary product, the blades, for a relatively higher price. Later, it was the printer. While printers are relatively inexpensive, the ink cartridge refills are quite expensive in comparison. This business model then evolved into a combination of product and service complementary offerings. For example, one can obtain a mobile phone for a relatively low price, but then add in the monthly service, and the cost can be astronomical in comparison. Ryanair evolved this complementary concept into a service-to-service model by offering airline seats for a very nominal price, but then charging for each ancillary item, including fees for handling credit cards, checking bags, priority boarding, etcetera. Ryanair is also reported to have looked into offering gambling on board, which could be another "side door" to making money.

The emergence and astronomical growth of the Web has yielded a whole new twist on the business model of making money through a "side door." For example, Chris Anderson states in his article in *Wired* magazine, that a business could drop the price of an actual product or service, rather than shift the cost of a complementary product and/or service, as illustrated in the above examples.[9] This shift in the business model can directly be attributed to the Web. Specifically, the Web can allow companies to offer services for literally free, and then make money elsewhere. As Jeff Jarvis, author of *What Would Google Do?* notes, Google is a great example of a company that illustrates the concept of "free." Jarvis goes on to give specific examples.[10]

- When Google purchased Blogger, it stopped charging for the service, but added advertising.

- When Google launched Gmail with its offering of lots of storage, it did not charge for the service, but had targeted advertising.

- When Google started offering directory assistance through 1-800-GOOG-411, it was free of charge, while most other mobile service providers were charging for each inquiry.

With respect to Google's directory assistance offering, Anderson projected that by 2012, Google could yield US$144 million by charging for directory assistance. However, he also calculated that by foregoing that revenue, Google could make US$2.5 billion in the voice-powered mobile search market instead. This example illustrates Jarvis' point that the successful ones follow this new trend, rather than those who are grasping at protecting old revenue streams and assets.

Anderson shows that there are numerous means of making money from the "side door." Specifically, Anderson breaks down the "free" economy into six categories.[11]

1. Users of the basic version of the subscription or service receive it for free. If a user wishes to have an upgraded version, she can do so, for a price. For example, Flickr, a photo sharing social network, is free (you can upload 100MB worth of photos each calendar month), but Flickr Pro (unlimited photo uploads, unlimited video uploads, unlimited storage, etcetera) costs US$24.95 a year.

2. Web-based ad formats, such as Google's pay-per-click text ads and Yahoo's pay-per-page view banners, evolved into more targeted advertising, such as paid inclusion in search results. Finally, the next steps include companies implementing pay-per-connection on social websites, such as Facebook.

3. Consumers get a product or service for little or no cost, but then pay a significant amount for the complementary service or product (as discussed earlier).

4. Consumers get the product at no cost. A good example is online music; some musicians offer their music online for free in hopes of promoting their concerts, paraphernalia, etcetera.

5. The mere act of using the free service yields value elsewhere, whether it be generating information or even improving the service itself. For example, when individuals use Google's 411 service, it enables the company to obtain valuable information, such as providing different accents which help Google to improve its service offering.

6. The Web has provided a forum where individuals can share and exchange on a global level. Two examples are (1) Wikipedia, where users generate, amend, and update the content, and (2) Freecycle, in which individuals can obtain secondhand goods for free if they are willing to take them.

Finally, another possible area of business model change is the use of games as an innovation tool. David Edery and Ethan Mollick show that games have become a powerful tool that organizations may use to recruit potential employees, teach employees, and motivate employees. While games have been used in the advertising and marketing world for numerous years, using games in this alternate arena is a relatively new concept. Why do games work in this manner? Edery and Mollick claim that games are attractive because they are able to target the very root of what makes people think, cooperate, and create. For example, in the corporate sector, Microsoft has used games to dramatically increase voluntary participation by employees in what used to be regarded as tedious endeavors, such as testing new applications for bugs. In the academic world, some medical schools have implemented game-like simulators as a means of training surgeons, which has decreased error rates. Specifically, Edery and Mollick recognize that games offer a means of teaching individuals three key skills that sometimes the traditional classroom setting has difficulty capturing.[12]

- Games can improve employees' ability to work in teams. It has been observed that the quality of teamwork can be a key indicator of the success of innovative projects. Specifically, it has been observed that if individuals feel comfortable communicating with one another within a team, the team has a better chance at learning faster, being more accurate, being more innovative, as well as working better with other teams. However, ironically, most companies do not focus much time or effort in getting the teams within the organization to work more efficiently.

- Games can teach individuals to utilize systems thinking. It has been observed that management often fails to understand the systemic effects of the decisions that they make. Games that teach systems thinking can be a tool to help managers to avoid and overcome such problems.

- Games allow people the opportunity to learn from a virtual experience when a real experience may not be available for reasons such as cost. Similar to the surgeon example above, medical residents can utilize simulators to learn the process of administering anesthesia. This practice is faster and the most cost effective means of completing their training. Furthermore, truck drivers who practice driving using games have been shown to have a decreased number of accidents as well as increased fuel efficiency.

Airline Examples of Business Model Innovation

Air Canada not only unbundled its product with respect to its pricing system (as mentioned earlier), but the airline has also unbundled its organizational structure. The idea was to create a holding company (ACE Aviation) with separate business units that could be viable on their own, that is, generate their own cash flows and financing, and provide reasonable returns to their shareholders. Air Canada then became a business unit under ACE. Examples of other business units include the monetization of Aeroplan (the frequent flyer loyalty program) and the spinoff of Jazz (the regional airline subsidiary). One key characteristic of this strategy appears to be that Air Canada's labor force no longer has access to the cash and resources of other business units, such as the Aeroplan.

In the case of LAN, it has developed a unique strategic alliance among four carriers (LAN Chile, LAN Peru, LAN Ecuador, and LAN Argentina) that produces enormous benefits for the members with respect to marketing initiatives, as well as savings in costs. For example, suppose LAN Chile were to operate a wide-body aircraft between Santiago and New York. It may not be able to maximize the utilization of its aircraft and crews if it were to maintain ideal departure and arrival times, and work around time zones, crew rest requirements, and so forth.

Now suppose, the aircraft starts in Santiago, stops in Lima, and then flies to New York (see Figure 4.3a). Next, the LAN Ecuador takes the aircraft with its own crew, flies southbound to Quito/Guayaquil, and then back north to Miami (Figure 4.3b). Then LAN Peru picks up the aircraft in Miami, and flies Miami-Lima, Lima-Miami, Miami-Lima (Figure 4.3c), Lima-Buenos Aires, Buenos Aires-Lima (Figure 4.3d), and, lastly, Lima-Miami (Figure 4.3e). Finally, LAN Chile takes the aircraft in Miami and flies it south to Santiago (Figure 4.3f), where the cycle can start again. Sharing the one airplane between three airlines provides the flexibility to maximize the utilization of the aircraft and crews to incredible levels. This example shows the coordination among only the three businesses of the LAN alliance. If one were to include the fourth member, LAN Argentina (whose network is also coordinated), LAN operations would achieve even greater flexibility to increase aircraft and crew productivity.

As discussed in the previous book, Google is far from being just another company, rather, it represents a whole new way of thinking and conducting business. Jeff Jarvis, in his book, *What Would Google Do?* applies this new way of thinking to a variety of industries, including the airline industry. Specifically, Jarvis takes what is often considered a commodity business, and reworks it under the Google principles, such as social networking. Jarvis begins with the idea that soon passengers will be connected in the air, just as they are currently on the ground, as airlines are finally beginning to implement in-flight wireless access. This accessory has the potential to totally change how passengers interact, such as through the emergence of passenger networks. For example, the flight could become a social experience, as it was once meant to be when planes included lounges and bars before they were replaced by revenue-generating seats. Not only could there be the social factor, but there could also be an economic factor as well. For example, passengers could be able to buy seats from fellow customers instead of only being able to buy them from the airline.

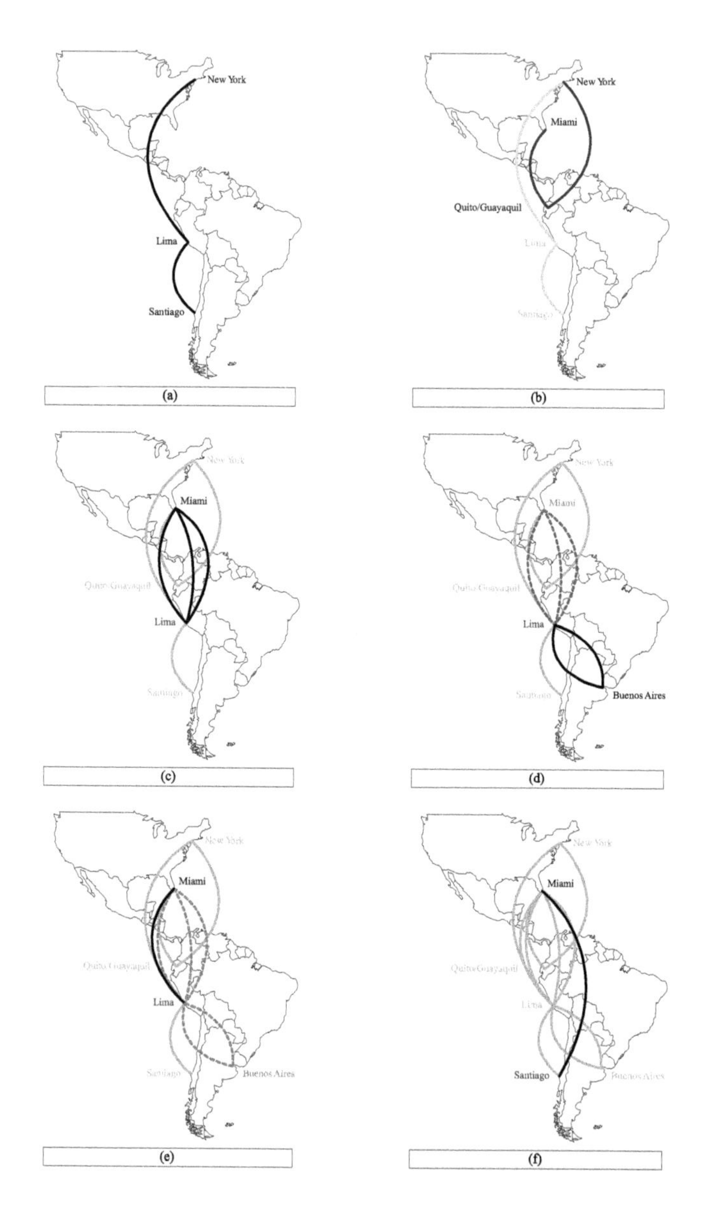

Figure 4.3 Example of Coordination Among LAN Partners
Source: LAN Airlines.

Case Studies in Innovation

Cirque du Soleil

One purpose of implementing an innovation is to enable the delivery of a value proposition. Finkelstein, Harvey, and Lawton give the example of Cirque du Soleil, the Canadian live entertainment company, which has created an extraordinary takeoff on the traditional circus. It continues to draw crowds year after year, decades after its first show in 1984.[13] Cirque du Soleil has grown from a group of street performers to a truly global brand built on creativity, with moving performances being offered everywhere from Buenos Aires to Tokyo, as well as permanent showcases in Las Vegas and Orlando. According to one reporter, Cirque du Soleil has "captured the public's heart." The company was ranked number 22 in Interbrand's 2004 poll of brands with the most global impact, beating out top brands Disney and McDonalds.[14]

Cirque du Soleil has achieved incredible success through implementing innovation. Specifically, Cirque du Soleil has taken some of the elements associated with the traditional circus and put it into a powerful story (each of the company's shows has a distinctive theme and story), described by music, choreography, and spectacular scenery.[15] Another unique characteristic of the company's shows is that they are continuous; there are no curtain drops to change the scenery. The scenes literally flow into each other with unbelievable transformations right before the audiences' eyes. Writers Babinski and Manchester recognize the company's incredible ability to blend creativity, imagination, and adventure into their productions. For example, elements, such as air, fire, and water, take on human form.[16]

The company continually creates and launches new shows, investing in innovation to ensure that its offering of entertainment stays new and exciting. They are literally constantly reinventing themselves.[17] The company utilizes driver number 3, leveraging competencies and resources, to expand its customer base in terms of both targeted customer segments and geographic location. For example, Cirque's *La Nouba* is a family-oriented show based in Orlando, while their *Zumanity* performances target adult-only audiences in Las Vegas. Since the premise of the show is music, movement, and scenery, rather than words, the performances are easy to adapt to global audiences.[18]

By implementing driver number 1, challenging conventional thinking, Cirque du Soleil has also been successful in terms of changing the cost structure and pricing model. Specifically, by eliminating one main element of the traditional circus, the animals, the company was able to eliminate a major cost. It also eliminated a major area of consumer resistance, namely, the animal rights lobby. Furthermore, by adding the element of theater, the focus and content of the event became more sophisticated, thus allowing higher ticket prices to be charged. Writers Kim and Mauborgne note that Cirque du Soleil has created a new and uncontested market space, a blue ocean of new market space, since the company is not trying to be a circus nor a traditional theater group, and, therefore, isn't concerned with the competition in either arena. The company has mastered a component of the blue ocean strategy by focusing on making the competition irrelevant by creating a totally new value proposition. This is what Kim and Mauborgne refer to as value innovation, and that is exactly what Cirque du Soleil did! They created a totally new offering, thus causing the company's value curve to stand apart. Most companies, especially those in the airline industry, tend to operate in the red ocean of bloody competition, where their company's value curves intertwine.[19]

Cirque du Soleil's success can be attributed to the company's focus on both constant innovation and delighting consumers. In addition, the company offers a unique experience that is not easily replicated.[20] The company recognizes the importance of not "becoming comfortable," as each show is one-of-a-kind, and it is this willingness to take creative risk that differentiates the company. This creative risk does come with a price. Each year, they invest over 70 percent of their profits back into the company in terms of R&D, new initiatives, and new shows.[21] Finally, in addition to the company's popular Cirque du Soleil shows, the company also has launched Cirque du Monde, a non-profit venture that travels into third-world countries, and teaches juggling, along with other types of imagination-building skills.[22]

Mahindra & Mahindra

The Mahindra & Mahindra Group (M&M), a family business in India since the mid-1940s, has become a global leader in the tractor and utility vehicle industries. Specifically, M&M ranks in the top three tractor manufacturers in the world.[23] In terms of the SUV market, the company introduced the Scorpio in 2002, which "took the market by storm." They also spent only approximately one fifth the amount that others in the industry would have spent. In addition, M&M plans to enter the US market with a hybrid SUV offering.

The company, which appears to have followed the economic history of India itself, did not place an emphasis on development or technology in the 1960s, as the company did not face much competition at that time due to India's closed economy. However, when the Indian economy opened up in the early 1990s, there was a big push for restructuring and change within M&M, including revamping business models and investing in technology. Despite these changes, the company did not perform well in the stock market, and was actually dropped by the Bombay Stock Exchange in early 2002. It was later re-added in mid-2007, because of CEO Anand Mahindra's effort to transform the Indian family business, which currently consists of approximately eight sectors, into a truly global leader.

Specifically, the drivers behind the turnaround of M&M include the CEO's identification of four key characteristics that he believes are the cornerstones of successful companies.

1. They aspire to be leaders in their businesses.
2. They have global potential.
3. They are innovative.
4. They display a ruthless focus on financial returns.

At a conference in 2002, Mr. Mahindra stated that all companies at M&M must conform to these principles. After one year, there was a dramatic turnaround, with profits and cash flow increasing in nearly every area of M&M. In 2004, the CEO announced the next of his "mantras," as he calls them, which challenged each area at M&M to become the most customer-centric in its industry. Record profits were achieved over a several year period as a result. In 2007, the CEO revisited the element of innovation, stating that innovation should become M&M's top priority, specifically, by building a customer-centric engine in each of the company's businesses.

The continual focus on innovation as the driving force in M&M's business model is partially due to the fact that the CEO believes that even once a company has successfully broken into an industry, it is still difficult to maintain that status and success, and innovation is the key to doing so. Specifically, five areas of innovation that the CEO of M&M focuses upon include:

1. Innovation has to start with *insights* about the customer.

2. Great products today have great *designs.*

3. Encourage *experimentation.*

4. Innovations must *add value* to the company's bottom line.

5. You need to have a *sales plan.* No innovation sells itself. Companies have to find ways of packaging and marketing it.

CEO Anand Mahindra sums up the above requirements of innovation in his acronym "IDEAS," which stands for insight, design, experimentation, added value, and sales plans for innovation.[24] It is evident that M&M's turnaround, and in turn, success, can greatly be attributed to the presence of driver number 4, supportive leadership.

Procter & Gamble

Nearly ten years ago, when A.G. Lafley was named (now former) CEO of Procter & Gamble, the company was in a state of crisis. Sales and earnings were declining. The company was running out of ideas for new products. P&G had tried to change too much, too quickly, which hindered them from focusing on running the daily operations properly. P&G needed clarity and focus in order to get back on track.

One of the key drivers in this revitalization effort was innovation. Until this point, P&G had been almost exclusively working on R&D internally. Specifically, P&G was obtaining less than 10 percent of its product ideas from outside the company.[25] Lafley recognized that the company could no longer continue this insular trend with respect to innovation. The insular culture had to be changed so that it would be open to communication and collaboration with outside innovators, customers, suppliers, etcetera, in order to obtain fresh ideas and opportunities. This required those working in product development and marketing to reframe their jobs. For example, those working in R&D could no longer experiment and invent alone in their labs. The new way would be outward-looking. This way, P&G's R&D division would be augmented by great scientists from around the world. Finally, the changed culture required that employees be rewarded for obtaining viable ideas from the outside in order to promote and reinforce such behavior.

One result from this change in mode of operations with respect to innovation is that relationships with outside vendors, such as suppliers, have become more collaborative rather than just transactional. For example, the focus is now on common business purposes, joint business plans, and, most importantly, on joint value creation. This new strategy is effective because it places the customer at the center of everything; the common goal is delivering better value to consumers.

This new customer-centric initiative was relevant as P&G was facing another challenge. The company was losing touch with its customers. Some researchers were more focused on the research itself, rather than upon how the innovation connected with the consumer. In response, the role of marketing was reinvented. Marketing and R&D partnered in the innovation process during every step, right from the beginning, in order to determine and match up consumers' needs with what could be produced. Another problem was that employees were spending all day in front of their computers and in internal meetings rather than out with the customers. The company learned that it could not successfully create solutions to satisfy consumer needs and fix customer problems without first gaining a deep understanding of these needs and problems. One means of gaining such knowledge is to spend time with consumers in their environments; homes, workplaces, etcetera. By implementing driver number 2, observing and experimenting, employees could gain an understanding of what kinds of products and services would improve the lives of their customers. This method of social anthropology, as discussed in the previous book, has been particularly successful in unveiling unarticulated needs which are discovered by observing the customer rather than just through discussions, as in traditional focus groups.[26] Now, not only do employees spend time observing customers in stores and in their homes, but consumers are spending time working in P&G offices with employees.

Now, P&G views innovation as a strategy and therefore incorporates it into every part of its organization; innovation is a part of its culture, its leadership, and its processes and systems. This includes the selection and implementation of the appropriate metrics, recognition, and rewards to encourage innovation. Innovation has now become the common thread of the company; it is at the core of P&G's business strategy. P&G has broken out of its insular ways, crossed geographical barriers, and gone so far as to invite anyone, but especially its customers, to get involved in the company's innovation process. Specifically, the company has implemented the "Connect + Develop" initiative to reach out to, and connect with, individual inventors.

This new approach to innovation has been a success. Five years after Lafley's new directive, P&G's innovation success rate had doubled.[27] Due to the shift from an internally-focused company to a collaborative one that embraces outside ideas, P&G has regained its stature as one of the most innovative, packaged-goods companies. P&G was recently rated number 12 in *BusinessWeek's* Most Innovative Companies (April 2009).[28] Not one airline made the top 25.[29]

Nintendo

Nintendo, the Japanese video game manufacturer, radically changed the landscape of the electronic entertainment industry with its Wii (pronounced "we") offering, which was launched in November 2006. At that time (in 2006), each of the three major game console manufacturers, Microsoft, Nintendo, and Sony, were all introducing new offerings to the marketplace. Microsoft and Sony followed the traditional path within the industry, namely improving graphics performance in an attempt to appeal to the "hard-core gamer."[30] However, Nintendo took a different path. First, Nintendo totally redefined the game console with the Wii. No longer is it a traditional joystick, but rather it is a sleek wireless wand, much like a remote control device for a television. Second, it contains a motion sensor that detects movement and rotation in three dimensions. The Wii console allows users to move around to play games, thus truly becoming a part of the game. For example, one can play a game of baseball by standing and holding the remote like a baseball bat, or one might serve a tennis ball by circling her arms overhead. Third, it is relatively easy to use, thereby capturing more of the marketplace by appealing to mainstream users. Specifically, the Wii has become popular among families and females, both nontraditional segments of the gaming market. This strategy of simplicity and ease of use was the intent of Nintendo to ensure that anyone, regardless of age, gender, or game experience, could understand and enjoy the product. Nintendo's innovation is an example of driver number 1. It challenged conventional thinking, as it truly changed the video game experience and target audience. It moved away from (a) competing on the traditional factors of graphics quality and game-play intensity, and (b) the traditional game itself, played by moving one's thumbs on a gaming console while seated in front of a screen, to giving players the ability to physically interact as well as socially engage.

Nintendo achieved success by following the strategy of design-driven innovation, a form of innovation that promotes aspects of innovation beyond technology alone. Nintendo effectively combined a radical innovation of substance with a radical innovation of technology. In terms of substance, it has redefined playing video games: active entertainment and physical exercise using the entire body for people of a variety of ages, rather than a passive, seated game in which a player only utilizes her thumbs. In terms of technology, microelectromechanical systems (MEMS) accelerometers enable the game console to sense the speed and orientation of the controller. MEMS have been used in the auto industry, such as in airbags, to detect if a vehicle was undergoing a serious accident. MEMS has also been used in the computer industry, to detect if a laptop is falling, and having such an event trigger a lock that secures the hard disk to protect it from damage. MEMS had even been used at Nintendo prior to the Wii, but then it was still just the old experience; the game remained the same in that movement occurred inside the game instead of with the actual player. What sets the Wii apart is that it does not just add a new functionality, but it creates a radically different meaning that is captured by all aspects of the product, the brand, the advertisements, and even the name (we infers the people who are actually playing the game).

The Wii has surpassed its competition, namely Microsoft's Xbox 360 and Sony's PlayStation 3. Specifically, in the first few months, it sold a million units. In its first six months, the Wii's sales soared, greatly surpassing those of Microsoft's Xbox 360 and Sony's Playstation 3 in terms of the US market. It is important to note that Nintendo's Wii was relatively less expensive than its competitors' offerings. Interestingly, in response to Nintendo's success with the launch of its Wii offering, both Microsoft and Sony announced price cuts to their respective game systems, and actually lost significant dollars on each console, according to estimates conducted by analysts. Ironically, while the Wii was less expensive than its competitions' offerings, Nintendo actually made a substantial profit on each unit. Nintendo was able to yield such a significant profit because the Wii contained relatively less powerful components, therefore costing less to manufacture. Finally, during the year following the introduction of the Wii, Nintendo's stock price rose significantly. A year after the release of the Wii, Nintendo introduced a new set of motion-sensitive devices, such as the Balance Board, that would allow a player to perform a myriad of balance games and exercises. Examples of balance games include spinning a Hula Hoop and skateboarding, while exercises include yoga movements. This new release in November 2007 augmented the Wii's meaning aspect even further, as it shifted it from a gaming entertainment system to more of a fitness tool. For example, a health club chain in Canada announced that it planned to implement the console in its exercise equipment, while a rehabilitation unit of a US hospital implemented the Wii to assist in the recovery of stroke victims.[31]

Allegiant Air

Within the airline industry, Allegiant Air would be an example of an innovation of the traditional business model. Although it is not a new business model, as it is based on the model of charter airlines based in Europe, it has been modified in the area of ancillary revenue. It provides a differentiating value proposition for its customers, and achieves profitability for the services provided. Out of the three, traditional, passenger segments, business, leisure, and visiting friends and relatives (VFR), Allegiant focuses strictly on the leisure and VFR segments. For the leisure segment, it provides extremely low fares to stimulate the market. It also relies heavily on the VFR segment, in that, although this segment is price sensitive, it is also relatively resilient. It does not pursue the business travel segment, and in order to penetrate the leisure and the VFR segments, it has developed an unusual network and a set of schedules, a heavily ancillary-revenue-based pricing policy, and an unusually low cost structure to offer low fares while still earning a reasonable return on investment.

First and foremost is the difference in Allegiant's network, which is truly a key differentiator. Their strategy is to link passengers in small, underserved cities to major leisure destinations, such as Las Vegas, Nevada, Orlando and Tampa in Florida, and Phoenix in Arizona. Las Vegas is connected to about 40 cities, Orlando to about 30, Tampa to about 20, and Phoenix to about 15. All routes are served point-to-point. The airline has tapped into a different segment of the marketplace; there are virtually no routes that compete with existing carriers. Frequency is very low. The average frequency on a route in its network (about 130 markets) is four flights per week. About a quarter of the markets served have only two flights per week.

Departure times are set to maximize load factors (over 90 percent) as opposed to aircraft utilization, and scheduled capacity is very flexible due to the very low investment cost of their airplanes (old MD-80s).

While all airlines in the US have implemented an ancillary-revenue-based pricing policy to varying degrees, Allegiant Air seems to have taken this practice to the furthest extent. It charges a fee for making a reservation not only through its call center, but even for reservations made through its website (a convenience fee). There is also a fee (for insurance) in case a passenger wants to make changes. Consequently, the percentage of fare derived through ancillary revenue sources is much higher than those of all other US airlines and it appears to be even higher than Ryanair, the acknowledged leader is this area. The airline also sells third-party products, such as hotel rooms, car rentals, and entrance tickets for shows and events.

Google

Google invented a new model in terms of organization and management. Earlier in this chapter, Google's economic model was discussed, but consider Google's "people" model. Specifically, consider how the original leadership at Google has overcome barriers that can often inhibit innovation by finding new ways in terms of people management, metrics, and monitoring. Interestingly, there is also innovation in how Google sources innovations.

First, Google's unique leadership structure has enabled the company to avoid such pitfalls as arrogance, egotism, and hubris that have plagued other companies and industries, such as discussed in the chapter on the auto industry (for more on the Auto Industry, see Chapter 2). Specifically, rather than having one leader at the top, Google has instilled a triumvirate (Eric Schmidt joined Larry Page and Sergey Brin in 2001). While this approach is atypical, it has been successful for Google. Why? For one, this leadership structure allows the others to "apply the brakes" to an inflated ego, for example, thus fostering a culture of mutual monitoring. Second, Google's leadership structure is conducive to a collaboration of expertise and perspectives. Third, such a structure allows errors to be corrected more quickly, as a single leader may be hesitant to change her decision, even if it is clearly incorrect, for fear of losing her position, or ego, or both. In the case of a triumvirate leadership structure, all three parties involved share responsibility.

Moreover, Google offers its employees more than merely financial motivation. Specifically, Google has worked to appeal to intrinsic motivation. One way the company has done so is by implementing a program similar to that of the 3M company in which employees are encouraged to devote a percentage of their work time (15 percent at 3M, 20 percent at Google[32]) to personal research projects that interest them. Thus, the employees are granted the freedom to explore and experiment, while the companies have created another source of potential innovations. The 20 percent policy at Google has directly yielded new products, such as Google Suggest, AdSense for Content, and Orkut. This work time strategy offers several benefits. First, it makes the company an attractive candidate as a potential employer to recent graduates who are transitioning from academia to the corporate world in terms of preserving some of the advantages of school, such as having time and freedom to explore during work hours. The strategy also augments productivity, as employees are motivated to complete their assigned projects in a relatively more efficient manner to be able to work on their personal projects. This is not to say that assigned work is rushed through, for Google's culture of quality insures otherwise. Hewlett-Packard has also implemented a similar program.

Third, while everyone involved in the tech industry knows that innovation is essential, and Google is certainly no exception, how Google is different is that it understands that innovation for the sake of innovation is not enough to shield a company from competition. Great companies, such as Kodak, Polaroid, and Xerox, could once preserve their position as market leaders with patents, but now, patents are not enough. Companies that want to surpass the competition and build a dominant position must utilize other methods, as discussed in the third driver of innovation, leveraging competencies and resources. How? One method is to increase the pace at which new products are released; Google seems to follow this path with its constant flow of new products and features. One unique element of Google is that it is not constrained by traditional R&D procedures. Therefore, ideas may be quickly transformed into products. Google focuses on listening, as the company knows that great ideas can come from anyone, anywhere, at any time. Google believes that innovation is everyone's business; Google's culture values original ideas from any employee. Therefore, it is not surprising that Google created a physical setting conducive to such idea generation. For example, the grand staircase in their main lobby has been adorned with electrical outlets so employees may sit on the steps and share their work with fellow employees.

Another means of obtaining ideas is through contests, such as the code competition that Google holds every summer (since 2002). This contest truly generates innovative ideas on a diverse and global level. For example, in 2006, Google received more than 3,000 entry applications, of which 630 were from students at 456 universities, and students from 90 different countries were represented.[33] Third, Google obtains ideas from its users. Fourth, Google buys ideas. It is important to note that Google acquires for the purpose of innovation, not consolidation. Specifically, rather than reinventing what already exists, the company instead purchases tools when appropriate. YouTube is one example. Google spent US$1.65 billion to acquire YouTube in October 2006.[34] YouTube had been first to market, and therefore YouTube possessed a valuable asset, as it already had a committed, massive, and growing user base. Thus, by acquiring YouTube, Google gained access not only to subscribers, but also to information on the behavior of those subscribers.[35]

Google clearly leverages multiple facets of the four drivers of innovation, from constantly experimenting in order to provide fresh offerings, to providing a setting conducive to experimentation and exploration, to providing supportive leadership. The entire Google culture promotes and fosters innovation. Only a handful of companies have been able to achieve such a culture.

Zipcar

Zipcar, an urban car-sharing service, was founded in 2000 in Cambridge, Massachusetts. The general premise behind the offering is that any member could reserve a car, walk to the location of the car, swipe her membership card, and drive off. This eliminated the hassles and costliness of ownership and parking in urban areas. The process, which has been compared to a condominium timeshare or airplane fractional ownership, only takes minutes to complete (pay a membership fee, reserve the vehicle online, walk to the car parked near your office, home, or dorm room, and then get access to the vehicle with one's Zipcard). Zipcar rentals only cost around US$10 per hour, and even include insurance and gas (a gas credit card is provided in the car for members to use). The concept, which is in alignment with driver number 1, challenging conventional thinking, was well accepted by the company's "model" customers: young, well educated, tech savvy, urban dwelling individuals. However, the model had some challenges, including the following. First, after Zipcar was founded in Cambridge, Massachusetts, the offering was quickly expanded through the Boston area, as well as New York and Washington, DC. However, the company ran into problems when it ran out of capital to further expand after using up so much to purchase cars in the initial cities. Second, there were not enough cars within each city, so members were having to walk too far to the nearest Zipcar, affecting the value proposition of convenience. Third, most members were interested in Zipcars during the evening and weekend hours, so cars were sitting idle during the weekdays.

To address some of these issues, Scott Griffith became the CEO of Zipcar in 2003. While he was intrigued by the company's concept and believed that it had the potential to grow exponentially in terms of membership, he also recognized the need for a new business model, one that would engage people so intensely that it would change their lifestyles. In order to do so, Griffith eliminated the company's original approach, and replaced it with a more distributive, self-service model (the company only has 250 full time employees). Specifically, when Griffith joined Zipcar, the company had only one vehicle in each of its locations, which concerned members. What if another member (referred to as a Zipster) wanted to use the car at the same time? The answer being that the second member would have to go to a Zipcar further away, again eroding the value proposition of convenience.

Griffith took the bold move of increasing Zipcar's inventory of just a few cars in each market to hundreds. For example, he placed 100 cars in certain areas of New York. He recognized that the company would lose money at first, but this strategy would lead to determining what number of cars would result in a profitable operation. The areas were selected by targeting zip codes that had a high percentage of residents who fit the "model" customer, as described above. Second, Griffith upgraded and expanded Zipcar's offering in terms of selection of cars. At first, the fleet mostly included Honda Civics and VW Beetles. Now, the offering includes what the company refers to as "mood cars"—a selection of 25 models (everything from Mini Coopers to Mazda Miata convertibles to BMWs) to cater to members' needs and desires—practical, fun, etcetera. These two key changes to Zipcar's model strengthened the company's engagement with its customers by offering more convenience in terms of the locations where the cars were stationed, as well as more in terms of car choices for whatever their desire may be—a fun drive, a trip, a lunch meeting, etcetera. These new strategies allowed members to not only get a vehicle whenever and wherever, but also to choose the type of vehicle they wanted to drive.

Griffith also implemented further changes which enhanced Zipcar's engagement with its customers. For example, he made transactions simpler and more transparent. Specifically, he changed Zipcar's system of hourly rates plus mileage to a simpler fee structure of an hourly charge that included 180 miles per day, so that members could better anticipate and estimate their total trip cost before actually embarking on the journey. This change was the result of interviewing members when they returned their Zipcars. Furthermore, Griffith enhanced the company's offering by giving members a discount if they pre-committed to specified amounts of driving. Finally, in order to address the issue of Zipcars sitting idle during weekday time periods, the company teamed up with corporate customers to use the service in place of having a company car as one solution. He even instituted co-branding arrangements, such as the one between Zipcar and IKEA, in which IKEA sponsored 14 cars in Boston adorned with both the IKEA and Zipcar logos. Members were offered a discount (reimbursed to Zipcar by IKEA) to use these cars, and were even offered "Zipcar parking only" spots at the IKEA stores, all in an attempt to attract Boston Zipcar members to leave the city and visit an IKEA store. The new model has been a success, sales and members have doubled each year. Some cities such as Alexandria and Arlington, Virginia, both just outside of Washington, DC, are actually offering programs that cover the membership fees for residents to join such programs in order to help with traffic and environmental issues. Finally, Apple recently announced a new iPhone application for Zipcar, which will enable individuals to find, reserve, and even control Zipcars. One may be able to unlock the Zipcar from her phone rather than utilizing the membership card, as well as potentially control other features such as the horn directly from her iPhone. The Zipcar concept has certainly taken off; what began as a "counterculture" movement in cities such as Cambridge, Massachusetts and Portland, Oregon has now become totally mainstream, with a growth rate of about 30 percent a year. However, competitors are quickly trying to get in on the action; Hertz, for example, has launched Connect by Hertz, a pay-as-you-go car sharing service that is currently available at several locations across the US, including on some college campuses, as well as in the UK and France. This is certainly a good example of personal mobility.[36]

Daimler's car2go service is another take on the car-sharing concept. Specifically, Daimler's car2go involves renting Daimler-made Smart cars by as little as the minute (one can rent by the minute, hour or day). The company's website promotes a "completely new, flexible type of mobility." The premise is that whenever an individual requires a car, she may rent it on the spot, or may reserve ahead either through her mobile phone or the Internet. The program has been offered as a pilot in the city of Ulm in Germany and Austin, Texas. It has been reported that Austin was chosen due to its proximity to a university as well as its consciousness for the environment.[37]

Insights for Airlines

The global airline industry has always benefitted from improvements in technology in different areas. Until the 1970s, the industry depended heavily on innovation related to aircraft technology to deal with product changes. Then in the 1980s, there was some management innovation related to the development of frequent flyer programs, followed by revenue management systems. In both cases, these innovations, identified and implemented by airlines, were copied by other industries, such as hotels and car rental companies. Technological innovation continued in the 1990s with the introduction of regional jets and electronic ticketing. Finally, since the beginning of this century, innovation has progressed in the areas of lie-flat beds, all Business Class flights in long-haul intercontinental markets, special check-in terminals for Premium Class travelers, more sophisticated in-flight entertainment systems, check-in via mobile phones, the use of Internet services in flight, and, more recently, in the area of customer experience through transparency (as discussed in the previous chapter).

However, to keep up with more fundamental change, both change that has already taken place and change that is expected, more innovation at the incremental and disruptive levels is required. For example, because it is now acknowledged that customers are becoming much more demanding, innovative thinking is needed to adapt the business model to enable airlines to view this change as an opportunity (for instance, to differentiate), rather than a threat. Unfortunately, for some airlines, being insular and thinking that the airline industry is too different (partly because of the inherent constraints) can, and has, to some extent, hindered innovation. On the other hand, progressive airlines are beginning to look outside their industry for areas of innovation to develop a sustainable competitive advantage, and, as such, make innovation a priority. Progressive airlines are also beginning to use their business intelligence capability to identify viable strategies to respond to change. Business intelligence can show an airline management what is going on, help identify viable strategies, and then enable management to monitor the success of the strategies implemented (as discussed in Chapter 5). Within the airline industry, the heart of the business is an airline's network. While most airlines have some systems (with varying degrees of sophistication) to evaluate their networks on an on-going basis, progressive airlines are exploring much more innovative models, based on business intelligence systems to help management fine-tune their networks (a) on a systems basis and (b) in as close to real-time as possible. However, really progressive airlines are planning to go much further with the use of business intelligence and adopt the best practices of merchandising (discussed in Chapter 5).

As the marketplace becomes even more competitive and as consumers become even more demanding, innovation will play an even more important role in enabling organizational viability and profitable growth. Consider five innovations, as recognized by SITA, which would allow simpler and smoother travel for passengers (mobile devices, Web 2.0, near field communications, radio-frequency identification, and biometrics), as well as five technologies that will improve air transport industry operations (service-oriented architecture, collaborative decision making, cloud computing, radio-frequency identification, and virtualization). Some of these advances were discussed in the previous book, while the remainder will be addressed in Chapter 7 of this book, *Flying with Tailwinds against Headwinds*. These ten innovations have the potential to change the traditional boundaries. However, the challenge, as recognized by SITA, is to find a cost-effective model that enables the airline industry to benefit from these technologies.[38]

It is evident from the discussion in this chapter that the key to implementing successful innovation, regardless of the industry, is placing the consumer at the center of everything and by developing world class consumer insight. However, while most airlines state and believe that they are customer-centric, precious few actually place the customer as the driver in the center of the innovation process. Consumer demographics, consumer behavior, consumer prosperity, and consumer ability to differentiate are evolving so rapidly that airlines must innovate beyond consumer imagination. Consequently, innovation is becoming the primary critical success factor.

For insights on innovation, airlines can look to the experiences of numerous non-airline companies, including the four discussed in this chapter (India-based Tata Motors, Canada-based Cirque du Soleil, USA-based Google and Whirlpool) as well as a few airlines, such as LAN.

- In the case of Tata Motors, it developed a new product, the Nano, based on providing a solution for a segment of the market (families riding on scooters in inclement weather) and produced a product at a cost that consumers in the targeted market segment could pay and net Tata a profit.

- In the case of Cirque du Soleil, the company designed an extraordinary way to showcase the traditional circus. First, it blended creativity, imagination, and adventure to provide a truly moving experience for the customer. Second, it changed the cost structure (by replacing animals with human artists) and the pricing structure (by charging much higher prices for theatre shows).

- In the case of Nintendo's Wii, the company not only focused on the technology factor of innovation, but it also considered the substance factor, an element that many businesses fail to recognize. By changing the design of the traditional game console, the company was able to provide an offering that was far more user-centric, as it truly created a physically active experience for all users of the game. Furthermore, the new game console is simpler, allowing the company to widen its market share by attracting "nongamers," rather than compete with the traditional companies, such as Microsoft and Sony, who were offering better, more advanced equipment and more complicated games aimed primarily at "hard-core gamers."[39]

- In the case of Google, the company implemented innovative ways of managing people and sourcing innovation. The leadership structure eliminates inflated egos, fosters collaboration, and allows errors to be corrected. Employees are motivated intrinsically, and not just financially. Finally,

Google is not dependent for innovation on its internal R&D processes and procedures. It is willing to go outside its company and make acquisitions that add value with respect to innovation.

- In the case of Whirlpool, the company took the time and effort to make innovation a priority within its organization. Specifically, it partnered with an outside vendor to foster innovation within the company. Whirlpool ensured that its innovation programs incorporated such aspects as alignment with other systems within the company, consideration of consumer insights and trends in customer behavior, and a business plan for experimentation, as well as an opportunity to redefine the business model based on actual marketplace experience. Most of all, Whirlpool's innovation program included employees from around the world, as well as from a variety of functions within the organization, staying true to its vision of innovation from everyone, everywhere.

- As for some insights from airlines themselves, Allegiant Air has developed a profitable network with almost no competition, very low frequency, and a highly flexible system for adapting capacity to the changing marketplace, based on the experience of charter airlines based in Europe. Such a product value proposition coupled with a low cost structure and a pricing system supported strongly by the generation of ancillary revenue have produced consistently high profit margins.

As an industry, as mentioned in Chapter 1, airlines have not been successful in making adequate profits to cover their cost of capital investments and associated risk. However, given the structural changes taking place in the global airline industry, the need for innovation is greater than ever either to differentiate as a product brand or accept being merely a commodity product. Innovation can be related to products, for example, at the incremental level. For instance, even if the current reduction in premium travel in intercontinental markets is not permanent and returns to its previous level, it could be at much lower fares, say, Business Class transatlantic fares at 50 percent of the level just prior to the 2008/2009 recession. This development could cause airlines to rethink about more innovative ways to develop and market Premium Economy Class services, or consider how innovation can be related to the way the product is distributed, or the way passengers are processed at the airport, or the way they are re-accommodated in cases of irregular operations. The possibilities and opportunities exist, and are plentiful.

Another example of a product innovation could be the potential of an open marketplace relating to frequent flier miles. Specifically, the airlines could allow passengers to bid on frequent-flier seats, upgrades, etcetera with miles. Next, airlines could use the passenger networks as a means of capturing knowledge. For example, (during passengers' return flight) airlines could inquire about hotels, restaurants, shopping, etcetera that they frequented at their destination. However, in order to be able to implement any of these concepts, airlines would need to not only have access to sophisticated business intelligence systems, but also give control, respect, and organization to their passengers, so that the passengers may, in turn, self organize into social networks and marketplaces. Keep in mind that passengers also have value to give. But as Jarvis notes, they will not share their value if they are not valued by their airline which, in many cases, is currently the situation.[40]

Finally, an airline could think about a change in its business model at the disruptive level. What if a global, brand-named, existing online, travel agency, or a new one started by a world-recognized Internet-based company, begins to achieve significant control of customers? Recall how, in the air cargo industry, forwarders developed a control of customers by providing an all-inclusive logistics solution (making the airlines mere "truckers in the air"), and how integrators entered the marketplace with a value proposition that included door-to-door services. Under such scenarios, airlines would either need to seek innovation that would revamp their existing business models within the industry, or be challenged by outside innovators such as Amazon, eBay, Federal Express, or Google. While this chapter outlined the need to undertake innovation to deal with the "new normal," the next chapter provides some tools and techniques to achieve this goal.

Basically, there are three "allies of innovation."

- Process innovations that allow the production of the standard airline product (travel from airport to airport) much more efficiently, perhaps with a twist into something called "cheap chic," as performed by carriers, such as jetBlue and Virgin America; process innovation can also relate to the ability of an airline to change configuration of its Premium Class cabins quickly to match capacity with demand.

- Product innovations that complement the airport to airport experience by adding smart interconnections to related ground transportation (same ticket, same terminal, same product standards), allow better usage of travel time through ancillary purchases (downloading music, in-flight ordering of products delivered upon arrival).

- Process and product innovations combined that bring new service or network patterns, such as on-demand scheduling, combined with "last minute flexible fares" (passengers "stored" at home or hotel able to hop on airplanes at short notice).

Notes

1 Material in this section is based upon the following sources: Estrin, Judy, *Closing the Innovation Gap* (NY: McGraw-Hill, 2009); Kelley, Tom, *The Ten Faces of Innovation* (New York, NY: Doubleday, 2005); Skarzynski, Peter, and Rowan Gibson, *Innovation to the Core* (Boston, MA: Harvard Business School Press, 2008).

2 Patnaik, Dev, with Peter Mortensen, *Wired to Care: How Companies Prosper When They Create Widespread Empathy* (Upper Saddle River, NJ: FT Press, 2009).

3 Snyder, Nancy Tennant and Deborah L. Duarte, *Unleashing Innovation: How Whirlpool Transformed an Industry* (San Francisco, CA: Jossey-Bass, 2008).

4 O'Brian, Jeffrey M., "Amazon's Next Revolution," *Fortune*, June 8, 2009, pp. 68–76.

5 Rigby, Darrell, Gruver, Kara, and James Allen, "Innovation in Turbulent Times," *Harvard Business Review*, June 2009, pp. 79–86.

6 Larreche, J.C., *The Momentum Effect: How to Ignite Exceptional Growth* (Upper Saddle River, NJ: Wharton School Publishing, 2008), p. 52.

7 Johnson, Mark W., Christensen, Clayton M., and Henning Kagermann, "Reinventing Your Business Model," *Harvard Business Review*, December 2008.

8 Ibid.

9 Anderson, Chris, "Free! Why $0.00 is the Future of Business," *Wired*, February 25, 2008.

10 Jarvis, Jeff, *What Would Google Do?* (New York, NY: HarperCollins, 2009), pp. 78–9.

11 Anderson, op. cit.

12 Edery, David, and Ethan Mollick, *Changing the Game: How Video Games Are Transforming the Future of Business* (Upper Saddle River, NJ: FT Press, 2009).

13 Finkelstein, Sydney, Harvey, Charles, and Thomas Lawton, *Breakout Strategy: Meeting the Challenge of Double-Digit Growth* (NY: McGraw-Hill, 2007), pp. 220 and 222.

14 Tischler, Linda, "Join the Circus," *Fast Company.com*, July 2005, Issue 96.

15 Sheahan, Peter, *FL!P: How to Turn Everything You Know on its Head—and Succeed Beyond Your Wildest Imaginings* (New York, NY: HarperCollins, 2007), p. 154.

16 Babinski, Tony, and Kristian Manchester, *Cirque du Soleil: 20 Years Under the Sun* (New York, NY: Abrams, 2004), p. 267.

17 Tischler, Linda, "Join the Circus," *Fast Company.com*, July 2005, Issue 96.

18 Finkelstein, Sydney, Harvey, Charles, and Thomas Lawton, *Breakout Strategy: Meeting the Challenge of Double-Digit Growth* (New York, NY: McGraw-Hill, 2007), pp. 222 and 247.

19 Kim, W. Chan, and Renee Mauborgne, *Blue Ocean Strategy: How to Create Uncontested Market Space and Make the Competition Irrelevant* (Boston, MA: Harvard Business School Press, 2005), pp. 12–4 and 40–2.

20 Finkelstein, Harvey, and Lawton, op. cit., pp. 247 and 320.

21 Tischler, op. cit.

22 Sheahan, op. cit., p. 58.

23 Mahindra Corporate Website: http://www.mahindra.com/OurGroup/Overview.html

24 Stewart, Thomas, and Anand Raman, "Finding a Higher Gear," *Harvard Business Review*, July-August 2008, pp. 68–76.

25 Merrifield, Ric, *Re-th!nk: A Business Manifesto for Cutting Costs and Boosting Innovation* (Upper Saddle River, NJ: FT Press, 2009), p. 187.
26 Taneja, Nawal K., *Flying Ahead of the Airplane* (Aldershot, UK: Ashgate, 2008), p. 139.
27 Merrifield, op. cit., p. 191.
28 "The 25 Most Innovative Companies," *BusinessWeek*, April 20, 2009, pp. 46–7.
29 Material for the P&G case study was also derived from the following sources: Hoque, Faisal, and Terry A. Kirkpatrick, *Sustained Innovation: Converging Business and Technology to Achieve Enduring Performance* (Stamford, CT: BTM Press, 2007); Lafley, A.G., "What only the CEO Can Do," *Harvard Business Review*, May 2009; Lafley, A.G., and Ram Charan, *The Game-Changer: How You Can Drive Revenue and Profit Growth with Innovation* (NY: Crown Publishing, 2008); Landry, Edward, Moeller, Leslie H., and Will Waugh, "James R. Stengel: Ultimate Marketers," in *CMO Thought Leaders: The Rise of the Strategic Marketer* (Edited by Geoffrey Precourt and Published by Booz Allen Hamilton, 2007); Nambisan, Satish, and Mohanbir Sawhney, *The Global Brain: Your Roadmap for Innovating Faster and Smarter in a Networked World* (Upper Saddle River, NJ: Wharton School Publishing, 2008).
30 Anthony, Scott D., Johnson, Mark W., Sinfield, Joseph V., and Elizabeth J. Altman, *The Innovator's Guide to Growth* (Boston, MA: Harvard Business Press, 2008), p. 80.
31 Material for the Nintendo case study was also derived from the following sources: Anthony, Scott D., *The Silver Lining: An Innovation Playbook for Uncertain Times* (Boston, MA: Harvard Business Press, 2009); Anthony, Scott D., Johnson, Mark W., Sinfield, Joseph V., and Elizabeth J. Altman, *The Innovator's Guide to Growth* (Boston, MA: Harvard Business Press, 2008); Ewalt, David M., "Nintendo's Wii Is A Revolution," *Forbes.com*, 11.13.06; Kleiner, Art, "The Thought Leader Interview: Tim Brown," *strategy+business*, Issue 56, Autumn 2009, pp. 93–100; Sheahan, Peter, *FL!P: How to Turn Everything You Know on its Head—and Succeed Beyond Your Wildest Imaginings* (New York, NY: HarperCollins, 2007); Verganti, Roberto, *Design-Driven Innovation: Changing the Rules of Competition by Radically Innovating What Things Mean* (Boston, MA: Harvard Business Press, 2009).
32 Girard, Bernard, *The Google Way* (San Francisco, CA: No Starch Press, 2009), p. 64.
33 Ibid., p. 82.
34 Ibid., p. 84.
35 Material for the Google case study was derived from the book by Girard. See Note 32.

36 Material for the Zipcar case study was derived from the following sources: Champy, Jim, *Inspire! Why Customers Come Back* (Upper Saddle River, NJ: FT Press, 2009); http://jalopnik.com/5283415/control-zipcar-from-your-iphone; Keegan, Paul, "The Best New Idea in Business," *Fortune*, September 14, 2009, pp. 42–52; Naughton, Keith, "Can You Give Up Your Car? New Auto-sharing Services Bet that You Can," *Newsweek.com*, updated 9:31 a.m. ET, August 4, 2008; Zipcar corporate website: www.zipcar.com

37 Material for Daimler's car2go was derived from the following sources: http://www.businessweek.com/globalbiz/blog/europeinsight/archives/2008/10/daimler_tests_c.html; http://www.car2go.com/portal/page/about/about.faces; http://www.smartcarofamerica.com/green_smart_cars/smart_car2go_in_texas_/; http://www.wired.com/autopia/2009/03/mercedes-car2go/.

38 Peter, Jim, "Ten Technology Advances that Will Change Air Travel," *A New Frontiers Paper,* SITA, 2009.

39 Markides, Constantinos, *Game-Changing Strategies* (San Francisco, CA: Jossey-Bass, 2008), p. 178.

40 Jarvis, op. cit., pp. 182–6.

Chapter 5
Firing on All Cylinders to Stay Ahead

The previous chapter discussed the need for greater innovation, not just to survive, but also to thrive in the new normal. It also discussed the major drivers of innovation shown in Figure 4.1. "Observing trends and experimenting" was one of them. This chapter provides additional detail relating to this area, namely the need to elevate to a much higher level the activities in three capability functions, business intelligence, analytics, and agility. This chapter ends with an example of different ways of pricing airline services based on the merchandizing model.

Elevating Business Intelligence, Analytics, and Agility

The use of data warehousing, mining, analytics and business agility varies significantly within the global airline industry. Few airlines know a lot about their customers' spending patterns and behavior. A few others know more about their competitors as well as more about the best business practices within other industries. There are even a few who have developed the capability, in terms of organizational structure and corporate culture, to not only deploy advanced analytics to connect their insights on customer behavior to their business decisions, but also to make business decisions in a much more timely manner. However, the new normal calls for the rest of the airline industry to move up the ladder of business intelligence, analytics, and agility. Even the few airline groups that possess these capabilities only deploy them for a small group of their customers. However, airlines need to deploy these capabilities for all segments of their customer base. Technology is available for airlines of all sizes and in all stages of development to move to a higher level of sophistication to increase their profit margins by becoming much more customer centric. Figure 5.1 shows that it is possible for an airline to optimize its RASM-CASM gap across the full spectrum from one passenger to the alliance level through an improvement in its business intelligence, analytics, and agility capabilities. This chapter provides some insights in these areas of capabilities, some examples of best practices, and a framework for price-optimization within the airline industry based on some experiences from the retail sector.

As mentioned in Chapter 3, consumer behavior of new generations, such as, the Millennials, is likely to be very different from past generations, such as the Baby Boomers and Generation X. Moreover, the new normal is likely to produce even a greater

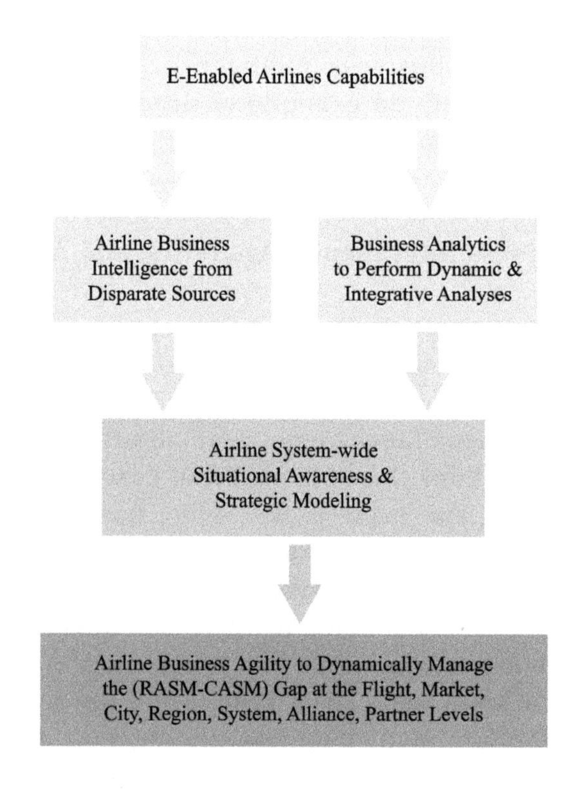

Figure 5.1 An Analytical Framework for Monitoring and Responding to Changing Customer Behavior

change in customer behavior. Chapter 1 showed the reduction in Premium Class travel resulting from the changed economic conditions, and hypothesized that these conditions may last a long time. The change in customer behavior can be the result of numerous factors, such as the advancements in mobile communications technology and the background and future impact of the economic downturn that began in late 2007. Even after recovery from this downturn, customer behavior could change significantly because of key developments, such as a change in the price of oil. Consider, for example, what the economic climate may be given different prices of gasoline, as author Christopher Steiner has done. Specifically, Steiner revisits consumers' reactions and behavior when gasoline hit US$4 a gallon in mid-2008, and then provides scenarios of what the landscape could look like as the price of gasoline increases by increments to US$6, $8, $10, and so on, as illustrated through his following scenarios:[1]

- *US$4*: Americans began to drive less. SUVs became less desirable, and consumers, at least in the US, started to realize that their lives must change. They also began to question the traditional way of thinking (bigger is better).

- *US$6:* Americans become deeply affected at this price level, and it signals the commencement of true change. While the SUV becomes greatly less desirable at US$4 a gallon, it disappears at US$6 a gallon. Diesel engines, not traditionally favored by the North American market, become more prominent. Americans drive even fewer miles (thus requiring less gasoline) and ultimately cause less federal fuel taxes being collected, which are needed for infrastructure improvements, such as road maintenance and mass transit systems. Public transit becomes a prominent method of transportation.

- *US$8:* This level leads to a dramatic change in the US airline industry. There will be fewer airlines, and the fares they charge will dramatically change consumers' flying habits, as trips to visit family and/or during holidays become less viable for the average consumer. Even wealthier travelers who frequent ski resorts and spas begin to spend less money on travel to such destinations, leading to a significant impact on the leisure destination/resort industry. Within the aviation industry itself, US aircraft manufacturers become affected by the overall lack of demand for air travel. Higher fares also affect other types of travel relating to such aspects as students' choice in attending universities.

- *US$10:* The role of gasoline-powered cars changes dramatically. They will still be in existence at this price level, but this price level will signify the end of a 75-year time period where the car defined, at least in the Western economies, consumers' identities and society in general. Vehicles used to be a statement of preference, style, and status, but will now become a mere costly utility. Electric cars become a way of life. Recreational vehicles (such as all terrain vehicles, snowmobiles, and jet skis) disappear, and family outings on a boat will no longer be economically feasible summertime activities.

- *US$12:* At this price level, there will be a massive shift of people residing in the suburbs to the downtown areas. There will be electric cars, but they may neither be cheap, nor plentiful. In any case, the transition to electric cars will take significant time.

- *US$18:* High-speed rail becomes prevalent in the US. Specifically, a network of high-speed trains connects cities, providing easy travel and seamless business. The US currently only possesses such a network in the northeast part of the country, but this will change (at this price of oil), as the US catches up with countries in Europe, as well as Japan.

While no one knows the price of oil in future years, such scenarios can be used to conduct "what if" analyses, using readily available business analytics that can connect insights on consumer behavior to decisions and actions.[2] These analyses, in turn, can lead to the development and maintenance of a competitive advantage through better and faster business decisions. In the case of airlines, they need to collect and analyze new data, data that is consumer behavior oriented. Let us examine, in some detail, the concepts of business intelligence, analytics, and agility, followed by the role that value and pricing can play in elevating the level of these concepts, and conclude with an airline application, specifically pricing airline services using a merchandising model. Examples from the experience of companies outside the airline industry will be highlighted throughout the discussion in this chapter.

Some restaurant chains are implementing business intelligence software in an attempt to obtain a better handle on consumer spending patterns. These patterns, in turn, assist management in terms of business analytics, making a myriad of business operations decisions, including everything from staffing levels to menu items.[3] Within the hospitality industry, Starwood Hotels & Resorts Worldwide is using business intelligence, with respect to the current trend of Baby Boomers moving away from luxury and toward "cheap chic" (made popular by brands such as Target, as well as IKEA, jetBlue and the Mini) for decision management purposes. Specifically, Starwood is implementing what they refer to as "style at a steal," which involves eliminating full-service restaurants, room service, and valets at their two, new, cheap chic hotel offerings: Aloft, named to reflect urban lofts, and Element, named to reflect a suite offering that addresses "elements" of a consumer's daily life. Starwood created a prototype of an Aloft and invited consumers to test the new offering. The test consumers found the changes, such as room service and valets (which were cut to help bring the room rates down) to be acceptable. Guests, for example, were fine with parking their own vehicles and exploring new restaurants as opposed to eating in their rooms. While some analysts think that Aloft may be a little "too hip" for Baby Boomers, Starwood has opened 25 Aloft hotels so far, at a rate of two hotels each month (one of the fastest growth rates in the industry).[4] These examples reflect the type of "ground-level intelligence," getting at the heart of consumer behavior and recognizing patterns and trends, which was referred to in Chapter 3, Customer Experience.

Business intelligence is also needed on the changing landscape so that leadership can not only face challenges by implementing effective defensive strategies to reduce vulnerabilities, but also to take advantage of opportunities by implementing offensive strategies faster than the competition. Business intelligence should show that there are new competitors. Consider an example from the hotel industry. If a top-tier hotel lowers its price enough, it can end up competing with a mid-level hotel, a scenario that did not occur in the old landscape. The entire landscape has changed; the Four Seasons is now not only competing against the Mandarin Oriental, but it is also competing against the local Travelodge. In the old world, this would have been unconceivable.[5] Similarly, British Airways is not just competing with Singapore Airlines in the London-Singapore market. It is also competing simultaneously with carriers as different as Emirates and AirAsia X.

A key need for airlines is to implement the process of connecting the business intelligence captured by the back office of the company directly to the frontline decision making, or business analytics, part of the company. The result will be greater business agility. Consider the following example of a luxury hotelier. A frequent guest arrives early, and her reserved room is not quite ready. The front desk employee is able to view in his system that the guest is not only a frequent visitor at that particular location, but is a valuable customer at the company's other properties. Even more important, he is able to quickly ascertain from the information provided in the system on the guest that she often utilizes the spa facilities. Instead of having the guest wait, the frontline employee, equipped with such information, is able to quickly offer the guest a complimentary spa service while the room preparations are being finalized. This transition from a possibly irritating situation to the customer to instead creating a "wow experience" opportunity is enabled by pushing intelligence from the back office directly to the frontline employees, who can use it to help analyze the situation, and make quick, on-the-spot, positive decisions, appropriate to the customer. This is certainly an opportunity to truly augment the customer experience.[6]

A key opportunity for airlines is the offering of on-board Internet access. While business intelligence indicates that this is an offering that many passengers will value, how much is the appropriate amount to charge for the amenity? A further question is the manner in which airlines will charge for this amenity, such as, by the hour, or by the duration of the flight? From a business analytics standpoint, has anyone done the analysis of what passengers are willing to pay for on-board Internet access? Predicative analytics is necessary to inform businesses, in this case, the airlines, of the probable outcomes if, for example, a certain program were to be implemented for this new amenity. Related to this question of business analytics, is the analysis of the consumer reaction to the implementation of change fees for business travelers. Southwest Airlines clearly did the analysis and ascertained that they should not, in fact, impose such a fee on their business passengers. This decision, in turn, has caused many business travelers to switch to Southwest Airlines for their business travel needs. Did the traditional carriers conduct this same analysis in terms of their business travelers before implementing fees? The level of passenger dissatisfaction, and its impact in the long term, suggests that they did not.

In dealing with the evolving landscape, top companies ask the question of how this recession is fundamentally changing their customers and their behavior, and adapt accordingly. For example, consumer electronics retailer Best Buy determined that one of their customer segments, home theater aficionados, was highly profitable and barely cut back their spending during a previous recession. Therefore, the company adapted by shifting resources to better serve this segment, and ultimately the company came out of the recession in a stronger position than that of the beginning. Furthermore, during the current downturn, Best Buy has continued to fare well, whereas its direct competitor, Circuit City, has closed all of its stores.

A different avenue of response to changing consumer behavior in the evolving landscape involves giving customers more control. An example of this method of adaptation is auto manufacturer Hyundai, who introduced a new value proposition to cater to changing buyer behavior in terms of uncertainty regarding the future and, therefore, fear of purchase commitment. Specifically, when most automakers were focusing on slashing prices during 2008 and 2009, Hyundai announced an alternate value proposition, one that had nothing to do with price. Hyundai's proposition stated that if a consumer bought a Hyundai, and then was unable to make the payments any time in the following year, the customer could give back the car, and the company would cancel the debt. Consumers responded favorably to the new value proposition. In the first month, when US sales fell 32 percent at Toyota, 40 percent at Ford, 49 percent at GM, and 55 percent at Chrysler, sales actually rose by 14 percent at Hyundai.[7]

Other companies, including some members of the telecommunications industry, are turning to social network analysis as a tool to better understand their customers, as highlighted in Chapter 3, Customer Experience. Specifically, while business problems in the telecom industry are often viewed through classical analytical methods such as analyzing call detail records, these methods often fall short as they yield results that fall under the "too little, too late" category. Rather, an alternate approach is to examine the social interactions that an individual might have with other people, and the influence that family members, friends, and business contacts have on one another. Social network analysis identifies key decision makers within such a network. For example, if an individual is a key communicator within her circle, then her choice to use one provider versus another may possibly have more impact on her network than if she were an idle participant in the network.[8]

In terms of cost savings, some businesses are reanalyzing their environment through a "green" lens. While a downturn may not seem like an appropriate time to launch an environmental initiative, as "greenness" may be viewed as a luxury, some companies are finding that reviewing their business operations from an environmental perspective can lead to additional savings that may have initially been missed. One example is Wal-Mart, which is renowned for cost cutting. One might think that the company would have already exhausted most of its cost savings opportunities, given the nature of its business model. However, when the company implemented an environmental initiative, it was as if they were viewing the business through a different pair of glasses, as one executive commented. The company found additional significant cost savings opportunities through implementing green initiatives in a variety of areas within the business. For example, by eliminating extra toy packaging, Wal-Mart found that it could yield a US$2.4 million savings per year in shipping costs, as well as saving 3,800 trees and a million barrels of oil. Furthermore, by installing auxiliary power units to control the temperature in the cabs of the company's fleet of over seven thousand trucks, drivers could shut off their engines during mandatory ten hour driving breaks (they used to have to let their engines idle during the breaks), yielding US$26 million in savings per year. Finally, by working in conjunction with its dairy suppliers, the Sam's Club division of the company was able to eliminate eleven thousand truck deliveries per year due to the implementation of a redesigned milk jug (as it fits more efficiently into the delivery trucks). All of these savings were revealed simply by viewing the business from a different perspective.[9]

While business intelligence can enable a business to ascertain trends in terms of the changing behavior of customers, and business analytics can help in the evaluation of alternative courses of action, it is business agility that calls for management to respond in a timely manner. This capability is becoming a necessity to conduct business in the global real-time economy in which there is hypercompetition and customers are well informed. Responsive companies find ways to address and meet consumers' unique needs and desires as they evolve. Most importantly, responsive companies act at once when they see an opportunity. They do not ponder for an extended period of time, as they understand that time is of the essence. One example of a company who has had success in the area of business agility and has set itself apart from others in its industry is clothier Zara. Zara has developed and implemented a highly responsive business model which only requires several weeks for its cycle time from development all the way to placing the merchandise in its stores; the traditional timeframe is closer to nine months for the industry. The company is responsive to changing consumer behavior, and the company's model involves factories and stores closely working together and coordinating to support the quick fashion changes that customers desire.[10] While Zara is an example of a company at one end of the business agility spectrum, GM is an example of a company at the other end. For example, GM's traditional planning bureaucracy was one factor that caused the company to miss out while Toyota and Honda jumped into hybrid technology, as also addressed in Chapter 2. Traditional, strategic approaches are not only time-consuming, they can also hinder the need to think and act fast.

The calendar-driven, formal strategic planning that has been embraced by most businesses is being reported to have become obsolete in today's environment.[11]

One way to be successful in the area of business agility is, of course, through the use of information technology (IT). However, leaderships of so many companies do not fully realize all the value that IT possesses, as business analysts Weill and Ross note. They argue that the successful companies in this digital economy will be those that are smart regarding how they utilize IT; IT should be a strategic asset, and if it is not, then it is actually a liability. Companies need to be IT savvy; the leadership must be able to use IT to increase performance on a consistent basis. It is important to note that businesses that are IT savvy do not have to be, and are not always, high-tech companies. While they can be high-tech, such as Google, they can also be traditional companies such as 7-Eleven Japan (convenience stores), Procter & Gamble, and United Parcel Service. Consider how 7-Eleven Japan utilizes IT to truly augment its operations. The company uses IT to empower its employees, all employees, right down to the salesclerks, with information. For example, every salesclerk has access to a handheld device to order stock. Salesclerks receive a constant flow of graphical data, which includes recent sales, product information, and even weather conditions. Equipping salesclerks with this type of data allows them to make educated estimations, such as how many hot noodles to order on a cold day. Moreover, the salesclerks receive feedback on the business results of their decisions, allowing them to constantly refine their estimations. This cycle affords the company the ability to continuously hone 7-Eleven Japan's inventory and revenue management decisions.[12]

How does a company move toward a responsive business operating strategy? One business writer addresses a few strategies for creating responsive business operations, the first being to automate the routine operations. He estimates that approximately 80 to 90 percent of activities within a business are routine, and these must be standardized and automated, stressing that it is here where an organization can focus on efficiency and lowering costs. Computers are faster and cheaper in terms of such work. Second, an organization must empower its people to handle the situations that are non-routine. Instead of bogging down employees with routine work, they should be given the tools (in terms of both data and training) needed to solve complex problems. Moreover, they need to be motivated. Training, motivation, and empowerment are the keys to employees being able to be responsive in complex situations. This is in alignment with Figure 3.1: The Three Drivers of Customer Experience in Chapter 3.

Centralized, hierarchical, pyramid leadership structures do not work in a fast-paced environment as the leaders are too far removed from the day-to-day points of action. Rather, responsiveness and agility occur when employees coordinate with each other to achieve a common goal. Employees are at the heart of the operations and see what is occurring, so provide them with the training and the motivation to respond appropriately. Companies that implement decentralized control structures, provide training and incentives for their employees to think and act for themselves, as well as equip employees with real-time performance data to make good decisions will succeed in outperforming the competition. Why? Employees who are engaged in self-directed, work-team environments with common objectives and goals will find a myriad of ways to implement improvements on a continual basis.[13] This is the case with grocer Whole Foods' highly empowered and self-directed work teams.[14]

Value and Pricing

While most businesses would claim to manage for value, value is a vague concept, as the term has different meaning to different people. There are different aspects that represent value to consumers, for example, price, convenience, dependability, image, integrity, and service. Normally, there may be only one or two aspects that are of real value to the customer. Other aspects may fall into the category of "nice to have." In the case of airlines, aspects perceived as valuable may include fare, schedule, frequent flyer program, seat comfort, in-flight service (food, entertainment, Internet access), access to airport lounges, self-service facilities, and re-accommodation during irregular operations. The question, then, is how to compute the value given to each of these attributes. One method could be to assign a benefit value to each of the value attributes. Another method could be to group attributes into a couple of broader categories, such as time savings (schedule, airports served, self-service services, and so forth) and brand (price, cabin service, customer experience, and so forth). The difficulty arises in that not only do different customers value different attributes differently, but also that the same customer may value different attributes differently at different times.

What is the experience of other businesses? Consider an example from the clothing industry, based upon the concepts promoted by business writer Dale Furtwengler. Brand, or in this case, image, is primarily associated with a designer's name in this industry. Attributes such as personal service and the store setting enhance the image value. Now consider the impact of image on clothing prices at certain stores. Wal-Mart, which offers no designer brands, also offers the lowest price. Target, which offers emerging designers, might price at two times the Wal-Mart price. Macy's, that offers designer brands and some added customer service, might price at four times the Wal-Mart price. Finally, high-end department stores such as Nordstrom and Saks Fifth Avenue, which offer all three image elements, namely designer brands, luxurious ambiance, and exceptional personal service, could command ten times the Wal-Mart price. This exercise illustrates the premiums that consumers pay for image; perhaps, as much as ten times the lowest price available.

How does such a pricing structure apply to other businesses, such as the airlines? To develop an image pricing model for a particular business, consider the following questions. First, determine how important image is to the consumers of the business. Are they in alignment with Target buyers? Macy's buyers? Nordstrom buyers? Second, consider the target customer, and research companies (in non-competing industries) that have similar customers. It is important to choose companies that have similar price points. Third, utilize the information from such other companies to determine the price multiples used at various price points. Next, determine which price points will be targeted, then utilize the highest prices multiple. If the offering is superior, it should be compensated for accordingly. Finally, to establish a price, multiply the above premium multiple by the base offering price within your industry (for example, in the clothing industry example, this would be the Wal-Mart price). This would determine the price of your offering. Such a system could work in an industry where image plays a role. Within the airline industry (which has thus far been considered to be a commodity business), this phenomenon is just beginning to emerge with the emergence of different categories of airlines.

Delving further into pricing, there is the concept of bundling. Bundling does provide several advantages. First and foremost, of course, it is an opportunity to increase revenue. Furtwengler gives the example of McDonald's, who can increase a sale by 48 percent by bundling a drink and hash browns in with the Egg McMuffin to offer a breakfast meal. This is an example of mixed bundling, as the components of the bundle are available either as a bundle or individually. He goes on to state that others can also experience average sale increases of around 40 percent through bundling offerings.[15] Bundling can also yield other positive outcomes, including boosting employees' productivity, and reducing risk. To illustrate these points, continue referring to the McDonald's example. Specifically, the employees are able to take more orders by not having to go through multiple questions, such as offering a drink with the order, offering a side with the order, etcetera. Bundling also reduces the risk of mistakes in terms of employees putting the orders together. Clothing retailers can bundle offerings such as shirts and ties together, to attract customers who value time, and also those who value assistance, that is, not having to worry about matching colors, etcetera, to increase average sales.

As author Rafi Mohammed points out, bundling is a pricing strategy that costs a corporation next to nothing to implement, but the strategy can have astounding results. Once again, think of fast food giant McDonald's. In 1997, McDonald's implemented bundling for a specific promotion, the Teenie Beanie Baby promotion. Mohammed recalls that during that time period, the full-sized Beanie Babies (collectible stuffed animal toys with bean bag filling) were extremely popular in the US. While there were hundreds of different types of Beanie Babies at that time, McDonald's commissioned mini-sized, or "Teenie," versions of the toy in ten different character offerings. However, one just couldn't walk into a McDonald's and purchase the Teenie Beanie Baby on its own for her child; one had to buy the Happy Meal, complete with a burger, fries, drink, for the special offering. Therefore, McDonald's Teenie Beanie Baby offering is an example of pure bundling, in which a cluster of products are only sold as a bundle, and only portions are available separately (the toy was only available through bundling). By implementing the price strategy of bundling to incorporate this special offering, McDonald's sales increased in three ways. First, there was an increase in the number of overall Happy Meals sold to the target market, children between the ages of three and nine. Second, sales of non-Happy Meal menu items increased as parents of the children purchased other menu items. Third, there were also adults who desired to possess the Teenie Beanie Babies, and therefore, they too, purchased Happy Meals.

Consider the following statistics surrounding the promotion. Prior to the Teenie Beanie Baby promotion, McDonald's was selling 10 million Happy Meals each week. However, nearly 100 million Happy Meals were sold in just 10 days when the Teenie Beanie Babies were added to the offering.[16] McDonald's Teenie Beanie Baby promotion is also a good example of business intelligence. The company dove deeply into its business intelligence system, and found an offering that not only appealed to their target audience for Happy Meals, children between the ages of three and nine, but also appealed to adults, thus greatly augmenting their sales of the Happy Meal offering.

Other industries, such as the hotel industry, have also used price bundling as a means of increasing revenue. For example, a hotel might bundle Internet access and use of the gym facilities together. Customers may select this bundle, even though they are not going to use the gym, but because they need to have the ability to check email from their room at night after a full day of business meetings. This may also be the case in the airline industry, specifically Southwest's Business Select offering. The bundle includes priority boarding, priority security lane access, and a drink, all for one price. However, a passenger may select this bundle just for the ability to get through the security and boarding processes more quickly, even though the passenger does not drink. Southwest has recently amended this offering by its new à la carte offering, launched in August 2009, in which passengers may pay an additional fee to reserve a spot in the boarding line, behind elite fliers, but ahead of families and other travelers.

Bundling also can provide an opportunity to lower costs. A good example is the auto industry. An automaker may only offer bundles of options, such as a sports package (perhaps a sunroof, leather seats, and a premium stereo), a cold weather package (perhaps heated seats, headlight windshield wipers, and backseats that split and fold to accommodate ski equipment), etcetera. This allows the manufacturer to reduce costs as there are not as many possible mixes of options that would need to be produced, thus reducing complexity. This strategy can also be advantageous to the manufacturer as more popular and less popular features can be combined, thus causing the consumer to purchase an option via a bundle that she otherwise would not have purchased. For example, a consumer may desire heated seats, and, therefore, must purchase the cold climate package, even though she did not need nor want headlight windshield wipers or split/folding backseats.

There is also the convenience factor of the price bundling strategy, as illustrated in the previously given clothing example (shirts and ties). The telecommunications industry has been extremely successful in positioning bundling as a convenience or a "one stop" solution to consumers' needs (one monthly statement for all services, one company to have to contact for service issues, etcetera). Since deregulation, corporations, such as the regional Bell phone companies who once offered only local phone service, can now offer bundles that include local, long-distance, and wireless phone services, as well as high-speed Internet. Consider the following statistics, although dated, regarding bundling in the telecom industry. In 2004, 62 percent of households bought two or more services from the same company, an increase of 20 percent in just two years (42 percent in 2002).[17] While price bundling has been successful in the telecommunications industry, it has not been such a success in the financial services industry. For example, Citigroup tried to implement the concept of one-stop financial services, but ended up selling its Travelers Life and

Annuity insurance company in 2005. One explanation for why bundling was successful in one industry, while not in the other, is possibly due to the fact that it is fairly easy for a consumer to change her telecommunications provider, while it requires much more effort to change financial services companies (for example, in terms of paperwork).[18]

Just as a company may decide to implement price bundling, they may also decide to unbundle their offerings. Dell Computer is a good example of a company that has unbundled its product offerings. Specifically, the company allows a consumer to build a computer online exactly to her specifications, thus truly customizing her purchase. They may literally go through a menu driven list of options, choosing exactly what they want at each step in terms of each option: choosing their system, the color of the system, the processor, the operating system, the software, the warranty and service, the display, the memory, the storage, the battery, and so forth. Also see Chapter 4 for another example of price unbundling, Air Canada.

Another aspect of price bundling is the consumption consideration. As John Gourville and Dilip Soman point out in their article in the *Harvard Business Review on Pricing*, many executives can recognize how pricing policies influence the demand for their companies' product or service offering, however, this is not so much the case when considering how pricing policies affect consumption (the extent to which customers use products or services for which they have already paid). Consider the example of a fitness club. Members may either pay each month or pay for a whole year at the time they sign up for an annual membership. The question is, which member is more likely to utilize the membership on a regular basis? Furthermore, which member is more likely to renew her membership when it comes due the next year? It would seem that they are most likely equal, given that in total, they have paid the same (US$50 a month versus US$600 for the year) for the same benefits. However, research indicates that the person paying by the month is more likely to go and exercise at the gym, because that member is continuously reminded of the cost of the offering as she receives a statement every month. At the same time, the person who paid for the entire year at the beginning may feel the need to get her money's worth early in her membership, but that feeling will likely fade the further away in time she gets from having made the payment. Finally, the member who has worked out more consistently throughout the year will be more likely to renew her membership for the next year, which is, of course, of key interest to the gym providing the offering.[19]

Leonard Berry, in an article in the *Harvard Business Review on Retailing and Merchandising*, stresses that prices are about more than just the actual dollars. There is also the psychological cost that is involved if a consumer feels that a business is being unfair in its pricing. For example, consider the psychology of a consumer concerned with whether an advertised sale price truly represents a markdown. Top retailers strive to minimize, or even eliminate, these psychological costs that can be associated with manipulative pricing. Furthermore, top retailers seek to follow the "everyday fair pricing" mantra rather than the "everyday low pricing" claim. According to Berry, many retailers leverage the "lowest price anywhere" position to make up for the lack of value-adding innovation. However, he goes on to recognize that retailers can, in fact, implement a fair pricing strategy if they are able to overcome two obstacles. First, they must make the cultural and strategic transition from the thought process that value equals price to understand that value is actually the total customer experience. Second, they must understand and implement the principles of fair pricing, such as offering most items at regular, but competitive, prices, and hold true sales periodically. Fair pricing practices also involve not having hidden charges, not taking advantage of temporary changes in demand by raising prices, and standing behind one's offering. These types of retailers recognize that while constant sales and markdowns on overinflated prices may increase sales in the short term, it is fair pricing, which, in turn, can win the trust of consumers that will yield the greatest pay off in the long term.[20]

Finally, how do you convey your value to the customer? As Furtwengler points out, one message certainly does not fit all. Rather, multiple messages are required, each containing a different, primary, value statement. For example, consider a scenario in which it has been determined that a company's customers will pay a premium for the following value attributes: friendliness, speed, and quality. Then the company's communications must reflect these consumer values. Specifically, one set of communications should focus on the value of speed, with friendliness and quality highlighted as add-on benefits. The next message would focus on the value of friendliness, with the other two as the added benefits, and so on. It is important to remember that the value propositions themselves are not being changed, but rather they are just being rearranged and emphasized in different manners in an effort to appeal to different buyers. Next, it is important to consider the language used within the communications. Some businesses utilize superlatives, such as "world class," to describe their offerings. Others go the other way under the premise that the offering will speak for itself. Neither approach is particularly effective, as buyers tend to be leery of companies who tout their own offerings, but at the same time, a company that does not get excited about its own offering, is most likely not going to excite others, either. Furthermore, some businesses focus on what they do or have, rather than what the consumer receives. For example, a health club may tout having "state-of-the-art" equipment and a trained staff, both of which are important, but what the customer is looking for is something that addresses their desire to lose weight.

Furtwengler lists the following points that must be considered in crafting the language of the communication to avoid the common mistakes listed above. First, the language must be customer focused. A good example is Enterprise Rent-A-Car's slogan, "Pick Enterprise. We'll Pick You Up," which certainly addresses what the consumer wants. Next, the language must be results oriented. For example, this can either be in the form of a testimonial by a customer, such as perhaps one who saved a percentage on her heating bill by the installation of the product, or if it is not possible to quantify, then use language that implicitly describes the experience, in this case, the result that the consumer can expect from the offering. Third, implement exciting language. This can be done in a myriad of ways, including by targeting buyers' dreams. Finally, the language must be appropriate for the value provided, as overstating value can cause feelings of distrust and skepticism.[21]

Airline Application: Pricing Airline Services Using a Merchandising Model

Numerous trends are driving a significant change in pricing products and services in the retail sector. These trends include a proliferation of retailers that are format invaders or provide "value" at one end of the spectrum, and dramatically change customer behavior at the other end. In between are trends, such as a blurring of retail channels. In the rapidly changing business landscape, retailers are moving to much higher levels of business intelligence and business analytics to optimize their pricing decisions. For example, the cost-plus pricing decision is at one end of the spectrum, requiring the least amount of sophistication with respect to an understanding and analyses of customer behavior. However, this pricing decision also leads to the lowest level of return. At the other end of the spectrum is customer-centric pricing. This approach requires a sophisticated understanding of customer behavior (at least down to a customer segment level, if not at an individual customer level), and the use of a comprehensive set of analytical systems to conduct, for example, what-if type analyses. The benefit, of course, is a much higher return.

Within the airline industry, pricing policies have varied from cost-plus (for charter airline services and feeder airline services) and competitor price match at one end, to customer segmented demand-based, at the other end. In the case of the latter approach, examples of customer segments are leisure and business travelers, and an example of the technique is revenue management. However, given the state-of-the-art business intelligence and analytics, as well as technology capability, it is now possible to implement customized pricing decisions down to an individual customer level. Up to this point, airline pricing decisions have reflected input from numerous areas: customer behavior, distribution channels, communication media, partners, and, of course, airline information systems (such as revenue management and loyalty). However, while most airlines do receive input from all of these sources, the information tends to be disjointed and more reactive than proactive. What is needed is a technology capability that makes the input interactive, not only among the three major constituents (namely, passengers, airlines, and their partners, such as hotels, and car rental companies), but also among the different functions within an airline (sales, revenue management, distribution, loyalty, and so forth).

Suppose a leisure passenger was inquiring about a fare to a leisure destination for a particular time period. Suppose, further, that the passenger indicated that the fare quoted was somewhat higher than the passenger expected. The agent could ask if the passenger also needed a hotel. If the answer was affirmative, the system could connect with a hotel's reservation system in real-time and could look into the room rates available in the type of hotel requested and the commissions that would be received by the booking airline. It may be possible for the airline to subsidize the airfare from the hotel commission, and make the sale possible. Similarly, the passenger could be asked if she was interested in tickets to a theater or a sports event that would be playing during the time period under consideration. Again, the commissions from these sources could be used to offset the lower fare desired by the passenger. For a business traveler, it may be even possible for the airline to negotiate higher commissions from a hotel, if the passenger was made aware of, and "persuaded" to utilize, some new features offered by the hotel, such as teleconferencing capability in the room. While, this is a form of dynamic packaging, it represents a much higher level of sophistication. It is in real-time, the airline booking agent would have access to the reservation system and inventory control of partners, and the offer to the customer would be based on proactive interaction.

It is the interactivity capability that enables the achievement of higher return pricing decisions through the fulfillment of customer needs (preferably at the individual customer level, or at least at the smaller customer segment levels). Let us go back to the retail sector for more insights, and consider buyers wanting to purchase a suit. Not only are there different types of stores, but there are also options within a given store, such as buying an "off the rack" suit, versus one tailored to meet a specific customer's needs. Even within the tailored category, there are options for selecting from a given set of designs (for pants and coats) to completely, individual selections. The closer the retailer, or, in our case, an airline, can get to a "one-off" interaction, the easier it is to shift from a price-driven to a customized pricing transaction. As mentioned in the example above, in the case of an airline, a passenger may not be willing to pay a particular fare, but is willing to include it with a particular price for a hotel with particular services. It is possible that the fees collected by the airline from the hotel may offset the lower fare demanded by the passenger. The deployment of customized pricing strategies can not only result in higher returns, but can also lead to greater loyalty.

Customized pricing strategies require a higher level of technology capability. Typically, an airline operates with multiple systems at the corporate level to meet the needs of different functions, commercial, operations, finance, and so forth. Even within one function, the technology department has multiple systems to meet the needs of one department. In the commercial area, there can be different systems to meet the specific needs of sales, revenue management, loyalty programs, distribution, and so forth. Even with respect to cases where each system could be categorized as "best of the breed," there is no true integration and interactivity among the numerous systems at any level; within a single function, among functions, or among alliance partners. While airlines have tried, and are still trying to develop integrated platforms, it is more realistic to seek technology that enables an airline to make its disparate systems interact among themselves than to throw out the multiple existing systems and acquire a single integrated system. The interactivity called for is not just within the different functions at an airline, but also among the different partners, may they be other airlines, distributors, hotels, or car rental companies. In other words, an airline cannot simply pass the customer from its website to the website of a partner. It must stay in the loop in real-time until the whole transaction is complete, making tradeoffs, where and when necessary to meet the needs of a customer while achieving high margins. Such a system would not only take dynamic packaging to a much higher level, but it would also put the control back in the hands of the airline. The latter decision-making feature will become a real necessity if traditional network airlines are to survive and thrive while competing with not only the increasingly expanding (and branded) low-cost airlines, but also the new global network airlines.

Insights for Airlines

It is commonly heard that of the many problems facing the global airline industry, the two most critical are overcapacity and lack of pricing power. It is also beginning to be heard that some of the more progressive airlines are beginning to look at the experiences within the retail sector to determine their applicability and viability for the airline industry, with respect to both pricing decisions and customer experience. Based on the experiences within the retail sector and the availability of technology capability to make customized pricing decisions, it would appear that airlines can at least tackle their key problems. Customized pricing decisions can lead to much higher return, greater customer loyalty, and an improvement in customer experience. Moreover, deploying the relevant, integrative, real-time technology capability means that airlines can integrate their operations more closely within the alliance structure without having to wait until common IT platforms have been developed. Consequently, it is possible to gain the benefits of consolidation through "virtual" mergers within the current alliance framework. If the concepts of business intelligence and business analytics (while not new) are revisited and moved to a higher level, airlines will be in a better position to face their on-going challenges, not to mention that they will be in a better position to take advantage of new opportunities.

The use of higher level business intelligence, coupled with higher levels of business analytics, can provide enormous value in other areas besides pricing strategies, namely, network, fleet, schedule, and product planning. Take the case of network planning. Airlines operating in international markets have historically used origin-destination traffic statistics from the global distribution systems. With increasing numbers of passengers carried by low-cost carriers (who have not used traditional global distribution systems) and the increasing number of reservations made via the websites of airlines, the historical origin-destination statistics may not reflect accurate information. Consequently, there is a need for better business intelligence and higher levels of business analytics to undertake network planning exercises. Similarly, with the significant decline in premium travel in long-haul international markets, airlines need to re-evaluate their aircraft seating configurations relating to the mix of First, Business, and Economy Class passengers. This decision raises, again, the need for much higher levels of business intelligence and business analytics.

Notes

1 Steiner, Christopher, *$20 Per Gallon: How the Inevitable Rise in the Price of Gasoline Will Change Our Lives for the Better* (New York, NY: Grand Central Publishing, 2009).
2 Rosenberger, Larry, and John Nash, *The Deciding Factor: The Power of Analytics to Make Every Decision a Winner* (San Francisco, CA: Jossey-Bass, 2009).
3 King, Rachael, "Business Intelligence Software's Time is Now," *BusinessWeek* (online), March 2, 2009.
4 Welch, David, "The Leaner Baby Boomer Economy," *BusinessWeek* (online), July 23, 2009.
5 Colvin, Geoff, *The Upside of the Downturn: Ten Management Strategies to Prevail in the Recession and Thrive in the Aftermath* (New York, NY: Penguin Group, 2009), p. 29.
6 Tobey, Bill, "The Dots: Driving Analytics to Front-line Operational Decisions Improves Bottom-line Results," *Teradata Magazine*, September 2009, pp. 34–39.
7 Colvin, op. cit., p. 108.
8 Bayer, Judy, and Edouard Servan-Schreiber, "Circle of Friends: Social Network Analysis Helps Telcos Better Understand Customers," *Teradata Magazine*, March 2009, pp. 18–9.
9 Colvin, op. cit., pp. 123–4.
10 Hugos, Michael H., *Business Agility: Sustainable Prosperity in a Relentlessly Competitive World* (Hoboken, NJ: John Wiley, 2009).
11 Greiner, Larry E., and Thomas G. Cummings, *Dynamic Strategy-Making: A Real-Time Approach for the 21st Century Leader* (San Francisco, CA: Jossey-Bass, 2009).
12 Weill, Peter, and Jeanne W. Ross, *IT Savvy: What Top Executives Must Know to Go from Pain to Gain* (Boston, MA: Harvard Business Press, 2009).

13 Hugos, Michael H., *Business Agility: Sustainable Prosperity in a Relentlessly Competitive World* (Hoboken, NJ: John Wiley, 2009).
14 For more on Whole Foods, see case study in previous book in the series: Taneja, Nawal K., *Flying Ahead of the Airplane* (Aldershot, UK: Ashgate, 2008), pp. 218–222.
15 Furtwengler, Dale, *Pricing for Profit: How to Command Higher Prices for your Products and Services* (New York, NY: Amacom, 2010), pp. 56–7 and pp. 102–3.
16 Mohammed, Rafi, *The Art of Pricing* (New York, NY: Crown Publishing, 2005), pp. 170–1.
17 Ibid., p. 170.
18 Ibid., pp. 169–170.
19 Gourville, John, and Dilip Soman, "Pricing and the Psychology of Consumption," *Harvard Business Review on Pricing* (Boston, MA: Harvard Business School Publishing, 2008), pp. 27–43.
20 Berry, Leonard L., "The Old Pillars of New Retailing," *Harvard Business Review on Retailing and Merchandising* (Boston, MA: Harvard Business School Publishing, 2009), pp. 1–18.
21 Furtwengler, op. cit.

Chapter 6
Viewing the Changing World Map

The Global Economy

The current combination of forces affecting the global airline industry is leading airlines to re-evaluate their networks, in some cases to reduce the number and type of markets served, and in other cases to expand the list of destinations served. In either case, the markets under consideration are being explored at different levels, ranging from individual city pairs (evaluated on the number of origin-destination passengers by fare type) to entire countries (those with large populations and growing economies), and, even at the level of continents.

Consider, to start with, the interest in growing economies. It was not too long ago when the argument went something like the following. While the 19th century belonged to Europe and the 20th century belonged to North America, the 21st century is likely to belong to Asia. Within Asia, two countries received the most attention, China and India. Then the general commentary began to be refined with indications that while the 21st century was still expected to belong to Asia (with China and India as the two major players), two other countries were expected to become markets of significant size, namely, Brazil and Russia. That led to the term BRIC countries.

More recently, it is beginning to be recognized that other growing economies with opportunistic markets for businesses are not limited to the BRIC countries. Growing economies are likely to be located all over the world, and not just within a particular region. First, *The Global Competitiveness Report* (2008–9) by the World Economic Forum that evaluates the status of more than 100 countries all across the world through a complex 12 component system, identified 15 countries that transitioned in economic development from stage 2 to stage 3, where stage 1 is classified as GDP per capita < US$2,000 and stage 3 is > US$17,000. While some of these countries are small islands (such as Barbados, and Trinidad and Tobago), or small in population (such as Croatia, Estonia, Latvia, Lithuania, and the Slovak Republic), or located in the energy-rich part of the Middle East (such as Bahrain and Qatar), there are a number of other identified countries with an estimated population of over 10 million (such as Chile, Hungary, Poland, the Russian Federation, Taiwan, and Turkey).[1] Second, other organizations, such as Forrester Research, for example, have also identified specific emerging countries, in this case, referred to as "The Next Eleven" or "N-11," that, along with the BRIC countries, have high potential of becoming the world's largest economies in the 21st century. Forrester's "The Next Eleven" include: Bangladesh, Egypt, Indonesia, Iran, Mexico, Nigeria, Pakistan, Philippines, South Korea, Turkey, and Vietnam.[2]

A third example would be an author, Vijay Mahajan, who focuses on Africa and identified four specific countries in Africa, Nigeria, Egypt, Kenya, and South Africa, known as the NEKS (pronounced "next").[3] A fourth example is the business researcher, Peter Marber, who is suggesting that there will be seven emerging markets, called the E7, that will have a population greater than five times the population of the G7 countries by 2050, and an aggregate GDP that is twice the level of the G7. Marber's E7 group is made up of Brazil, China, Indonesia, India, Mexico, Russia, and South Korea,[4] while the G7 countries include Canada, France, Germany, Italy, Japan, the UK, and the US. The fifth example is a recent expansion of BRIC to BRIICS that adds two more countries to BRIC, namely, Indonesia and South Africa.[5] Figure 6.1 shows the location of the BRIICS, NEKS, and E7 countries. It is important to note that in Figure 6.1, only the densely populated areas of China and Russia are shown. The function of the diagram is for illustrative purposes only, and not to show the accurate location or geographic boundary of any particular country.

It is in the preceding context that one could assume that the 21st century is not likely to belong to any particular region, but rather, it is likely to be referred to as a global century with high-growth and competitive countries (including most of the BRIICS countries) located all over the world. A dozen or so of these increasingly competitive countries have the potential to provide tremendous opportunities for not only airlines based in their countries, but also for other airlines to connect with those countries. This chapter continues with a brief discussion of four widespread and varied countries from around the globe, Chile, Egypt, Malaysia, and Turkey, and includes some opportunities for airlines based in these countries. The discussion then moves to the two regions, often ignored by the global aviation community, Africa and Latin America. One explanation for this lack of attention is that each continent accounts for only about 5 percent of the total global travel in terms of passenger kilometers flown. It could also be due to the difference in people's perceptions about the opportunities relating to these two continents, perceptions that, in turn, are most likely based on the general images portrayed in the media. The brief synopsis of paradoxes (containing both pessimistic and optimistic perceptions) relating to both continents will be followed by sections highlighting potential business opportunities as well as challenges and opportunities for airlines.

Figure 6.1 BRIICS, NEKS, and the E7 Countries

Four Examples of Growing Economies and Opportunistic Markets for Businesses

Chile

Chile has moved its position upward in terms of economic development (currently in transition from stage 2 to stage 3), ranking number 28 on the Global Competitiveness Index.[6] This country was also selected to be discussed because of its status as a role model for other countries within Latin America, which can be attributed, in part, to its execution of economically sound policies for nearly three decades. Chile has a population of nearly 17 million, of which about 85 percent live in urban areas, including 40 percent residing in greater Santiago.[7]

- Chile has maintained sound economic policies since the 1980s. These policies have contributed to steady growth. Chile's economy has been growing at a very healthy rate, with growth in GDP averaging almost 6 percent per year between 1990 and 2008. This rate is much higher than other major Latin American economies, such as Brazil, Mexico, and Argentina. It is, in fact, in line with a number of fast-growing economies in Asia (such as Thailand, Malaysia, India, and South Korea). The country's poverty rates have been reduced by over half (incomes below the poverty line fell from 46 percent in 1987 to around 18.2 percent by 2005[8]), and have helped secure the country's commitment to a democratic and representative government. It has been reported that Chile's "secret for success" can be attributed to a combination of prudent fiscal policies, a focus on reducing inflation, a freely floating exchange rate, and trade openness. Within the Latin American region, Chile ranks at the top in terms of ease of doing business, as government regulations are reported to be less burdensome in Chile in comparison to other areas of Latin America.

- Chile is known for its copper. The north Chilean desert contains great mineral wealth, especially copper, and the state-owned firm, CODELCO, is the world's largest copper-producing company. Chile has also made an effort to expand into other, nontraditional exports, including forestry and wood products, fresh fruit, processed food, seafood, and wine. At the same time, Chile has also been diversifying its economy by reducing the share represented by manufacturing and mining, and increasing the share represented by services. For example, the growth in air transportation services has been facilitating the government's strategy to move toward a more service-based economy.

- Chile's population is relatively well educated. A pool of well-educated computer engineers is attracting IT companies, not only from the US, but also from India.[9] The country also possesses a top-ranked healthcare system.

- Relations between the US and Chile are strong. The US government recognizes the importance of Chile's maintenance of a vibrant democracy and a healthy and sustainable economy. As a result of the 2004 trade agreement between the US and Chile, approximately 90 percent of US goods can enter Chile tariff free (tariffs for the remaining 10 percent are to be phased out by 2015). Chile has also entered into trade agreements (although not all of them are full free trade agreements) with a myriad of other nations, including China, India, Mexico, and South Korea, as well as the EU. Besides the US-Chile FTA, the two governments consult frequently on issues of mutual concern, including the areas of culture, diplomacy, science, and security.[10] Consumer tastes tend to be Americanized; for example, shopping malls are very popular, and the use of credit cards is increasing.

As for air transportation, LAN (Chile's flag carrier) is an example of an airline where management has a true opportunistic mindset to develop initiatives that work around constraints to capitalize on opportunities. Given the traditional regulatory constraints relating to airline ownership and control as well as the provisions of air services agreements (in a highly regulated continent), LAN has developed one brand with four production units (subsidiaries or affiliates) by acquiring airlines based in Argentina, Ecuador, and Peru. Figure 6.2 shows the extent of the network of the airline, with the main airline based in Chile. LAN not only has extensive domestic networks within each of its four units, but it has also developed an enormous international network by strategically combining the resources of each unit (traffic rights, fleet, crews, airport facilities, and so forth). In addition, LAN has developed an enormous cargo network, consisting of not only the services provided by these four subsidiaries, but also LAN cargo carriers based in Brazil, Colombia, Mexico, and the US. This diversification of the LAN network provides a significant diversification in its revenue base that is now approaching US$5 billion annually, and cargo operations generate more than one-fourth of this total revenue. LAN's model could easily be expanded to include passenger subsidiaries in countries such as Brazil, Colombia, and Mexico, and these types of initiatives could open up even more global opportunities. For example, direct flights to China from Lima, Peru, São Paulo, Brazil, and Mexico City may be possible, and such initiatives would also provide the holding company, LAN, the flexibility to schedule aircraft and crew to maximize productivity within the constraints of government regulations.

Figure 6.2 The LAN Network

Source: Information based on the information available on the airline's website.

Egypt

Egypt was selected for discussion because it is a country with a myriad of strengths, from its natural assets, ranging from the Nile River and the Suez Canal, to its oil and natural gas reserves, and to its tourism industry. While the country has been ranked as a stage 1 in terms of economic development, it has many strategic, as well as lucrative, attributes (as illustrated below), making it a source of great potential.[11] The country, occupying the northeast corner of the African continent, is the most populous country in the Arab world (with a total population approaching 80 million), as well as the second most populous on the African continent. Consider the following attributes.

- In terms of transportation opportunities, the Nile River is a key asset for local transportation, while the Suez Canal is a major waterway of international commerce, as it links the Mediterranean and the Red Sea, allowing ships to avoid the lengthy trip around Africa to pass from the Mediterranean to the Red Sea. The country possesses a well-maintained road network that has been rapidly expanded. Egyptair offers domestic air service to major tourist destinations from its Cairo hub, in addition to overseas routes.

- The Nile Valley and Delta provide great agricultural opportunities; approximately one-third of Egyptian labor is engaged directly in farming, and many others work in the processing or trading of agricultural products.

- Egypt's natural resources include petroleum, natural gas, phosphates, and iron ore. Crude oil is found primarily in the Gulf of Suez and in the Western Desert, while natural gas is found mainly in the Nile Delta, off of the Mediterranean seashore, and in the Western Desert. The country enjoys a high volume of tourism; over 50 percent of Egypt's economy is involved in the tourism industry. With investment in such areas as cleaning up the water pollution plaguing seaside towns featuring water sports as tourist activities, this percentage could be increased further.

- Egypt maintains a strong and friendly relationship with the US, partially due to a shared mutual interest in Middle East peace and stability. The US has a large assistance program in Egypt that provides funding for a variety of programs. In addition, US military cooperation has helped Egypt modernize its armed forces and strengthen regional security and stability. Egypt, in turn, has provided military assistance and training to a number of African and Arab states. In terms of the Arab community, Egypt is a key partner in the quest for peace in the Middle East and resolution of the Israeli-Palestinian conflict. In terms of the African community, Egypt contributes regularly to UN peacekeeping missions, most recently in East Timor, Sierra Leone, and Liberia, as well as in August 2004, where Egypt was actively engaged in seeking a solution to the crisis in the Darfur region of Sudan.

- China is forecasted to replace the US as Egypt's top trading partner due to its investments in such projects and ventures as cement factories, convention centers, electronics companies, and the Suez Canal. Furthermore, Europe is also moving to the top spot over the US in terms of its influence in Egypt. Specifically, while the US has shied away from Egypt's intricate bureaucracy, European companies, on the other hand, have spent millions of dollars to upgrade sectors including agriculture, construction, and tourism in Egypt. Such strategic advances by both China and European countries are augmenting Egypt's position in the geopolitical marketplace.

- Egypt serves as both the cultural and the informational center of the Arab world. Cairo is the region's largest publishing and broadcasting center.[12] However, as with a number of other developing countries, there are concerns relating to corrupt political party oppositions.

In terms of transportation opportunities, Egypt has always been a leading country in its region. For example, it was the first country in Africa to establish a railroad (around 1850), and it has been a leading African country in the development of transportation and communication systems. In terms of air transportation, Egyptair has a global network that provides extensive services within Egypt (eight destinations from Cairo alone) and within Africa (about 15 destinations), as well as to and from Europe (about 24 destinations), the Middle East and the Gulf (about 17 destinations), Asia and the Far East (four destinations), and North America (two destinations). Code-share flights serve additional networks. Given the geographic location and the population and economic base of Egypt, there are substantial opportunities for future growth. For example, in China, currently, only Beijing is served by Egyptair while Guangzhou is served by both Egyptair and a code-share partner. Given the growing relationship of numerous African countries with China, there are further network opportunities for air services to and from China, connecting through Cairo.

Malaysia

Malaysia was selected for discussion due to its powerful transformation from a mere producer of raw materials into an emerging multi-sector economy. It is listed as being at stage 2 in terms of economic development, and ranked number 21 on the Global Competitiveness Index.[13] The country has an interesting geographic composition: it is separated by the South China Sea into two regions, Peninsular Malaysia and Malaysian Borneo. Malaysia has a total population of over 25 million.

- Since Malaysia gained its independence, its economic record has been one of Asia's best. Specifically, the country used to be heavily dependent on primary products, such as rubber and tin. However, Malaysia has grown into a middle-income country with a multi-sector economy based on services and manufacturing. Malaysia is one of the world's largest exporters of semiconductor devices, electrical goods, and Information and Communication Technologies products. The country also possesses oil and gas for export. The country is a significant oil exporter in all of East Asia due to its diligent practices of leveraging technology, as well as managing its reserves wisely. Malaysian leaders have strived to advance the country's economy in terms of value-added production by attracting investments in high technology industries, medical technology, and pharmaceuticals. Specifically, the MSC Malaysia (also known as the Multimedia Super Corridor) is a government designated zone created to fast-track Malaysia into the information and knowledge age. MSC Malaysia contains companies which are focused on supporting Malaysia's initiative of promoting the global Information and Communication Technologies (ICT) industry.

- Malaysia possesses infrastructure that is on par with that of first world nations. For example, the country has constructed planned cities, Putrajaya and Cyberjaya, both located near Kuala Lumpur. These planned cities are part of the MSC Malaysia. While Kuala Lumpur continues to serve as the country's capital, Putrajaya serves as the federal administrative center of Malaysia, and Cyberjaya serves as the Silicon Valley of Malaysia. Furthermore, the Petronas Twin Towers in Kuala Lumpur are among the tallest in the world, complete with a double-decker sky bridge which connects the two at the forty-fifth story level.

- Malaysia and the US cooperate closely on security matters, including counter-terrorism, maritime domain awareness, and regional stability, with the militaries of each country involved in numerous exchanges, training, joint exercises, and visits. The US and Malaysia share a diverse and expanding partnership, including strong economic ties. Specifically, the US is Malaysia's largest trading partner, while Malaysia is the 16th largest trading partner of the US. Furthermore, the US is the largest foreign investor in Malaysia on a cumulative basis, with American companies being particularly active in the energy, electronics, and manufacturing sectors.

- Malaysia has strategically become friends with nearly everyone, just as other Muslim oil-producing states, including Saudi Arabia, have done. At the same time, Malaysia strategically assures the US it is on its side while also ensuring not to upset China. Malaysia pledges to remain neutral should any situation arise, much like other countries including Australia, South Korea, and Thailand.

- Malaysia is characterized as being conservative, while simultaneously open and tolerant. The country is a melting pot of many ethnicities, including a large Muslim population. It is interesting to note that Asia is home to most of the world's Muslim population, not the Middle East.[14]

- Each year the World Bank produces a report, *Doing Business*, which tracks changes to the government regulations that affect business. In the 2010 report, Malaysia ranked number 23 out of 183 (with rankings relating to the ease of doing business). It is interesting to note that in this report, Malaysia ranked higher than such economies as France and Germany.[15]

As for the air transportation industry, Malaysian Airlines has become a textbook example of how an airline can rise from near bankruptcy to a profitable business in less than three years. The problem was not related to the lack of opportunities for growth, given the potential growth of the country; rather it was related to "mismanagement" of the airline. Previous problems related, in part, to a number of unprofitable routes operated by the government-owned airline, presumably to protect tourism, and, in part, to the high number of employees relative to the size and operations of the airline. The turnaround started with a vision that was leisure focused and business oriented, and a four-pronged strategy that related to "dynamic pricing, network optimization, cost management and innovation." The cost element was particularly important, given the decision to focus on the leisure segment that is often price sensitive.[16] With respect to the long term, there are enormous opportunities for growth into the capitals of ASEAN countries, South Asia, China, Australia, and the US West Coast, as well as the Middle East, and Europe.

The other development to support the expansion of air services to and from Malaysia is the enormous expansion of AirAsia, based in Kuala Lumpur. AirAsia has not only expanded at an incredible rate within the region, but its affiliate, AirAsia

X, is expanding into intercontinental markets; London, in the UK and four destinations in Australia. Figure 6.3 shows, for example, how AirAsia X can connect its long-haul international flights coming from China and Australia to the UK via Kuala Lumpur. AirAsia X is now planning to establish a "virtual hub" in the Gulf region of the Middle East with services that would enable it to increase the utilization of its long-range aircraft. Consequently, the growth of the country and its two airlines are expected to be complementary.

Turkey

Turkey occupies a truly strategic location, lying between Europe, Asia, and the Middle East. With a population of over 70 million, Turkey is a large, middle-income country with an economy that is diversifying continually, and moving from agriculture to industrial products. Turkey's strategic location is also playing a major part in the flow of oil. The population is 99 percent Muslim, but in

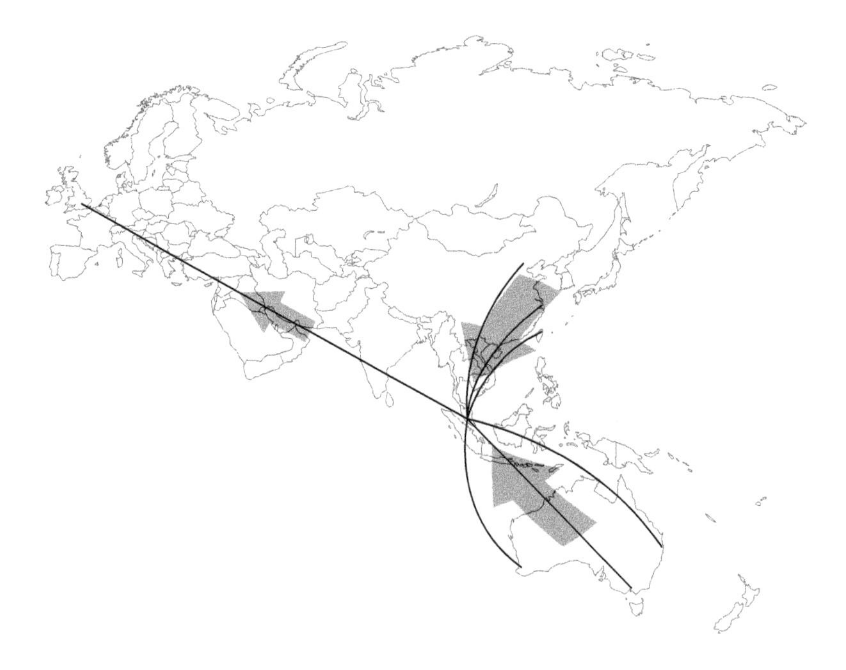

Figure 6.3 An Example of Intercontinental Connections by AirAsia X

the major cities, such as Istanbul, with the existence of secular principles of the Turkish Republic and multiple ethnic groups, one can see a transition to a Western society while maintaining the Muslim faith. Turkey is more like a European than an Asian country, even though only about 3 percent of Turkey's area is in the European continent. There is also an income gap between Turkey and Western Europe. As for tourism, Turkey is becoming a global player. It is becoming known as a center of tourist experience with historic sites, ruins, and treasures that rival Italy and Greece. Along with history, its modern attributes are evident from the existence of high-speed, fibre-optic communication connections, even in remote villages. Turkey has already become a major tourist destination in Europe and has the potential to be in the top one half of a dozen tourist destinations in the world.

Istanbul is an enormous metropolis where East truly meets West. In fact, Istanbul is even physically divided, with one part of the city in Europe and the second part in Asia. The two Bosphorus Bridges (the first being built in 1973) connect the two parts of the city and the two continents. Although about 90 percent of the city's population is Muslim, it is made up of numerous ethnic groups. Istanbul, with a population approaching 15 million, has become a great world city, and a cultural, financial, and historic center.

- Turkey is a founding member of the Organization for Economic Co-Operation and Development (OECD), a member of NATO (with NATO's second largest army), and, currently, a European Union candidate.

- Economically, although about one-third of Turkey's employment is still related to the agricultural segment and another one-third to the textile and clothing segments, the mix is changing rapidly through an increase in employment in the industrial sectors that relate to automotive, electronics, chemicals, and petrochemicals. Although the economy became unstable at the turn of this century, as evidenced by the collapse of the banking system and inflation running over 100 percent, the government implemented unprecedented reforms that not only brought economic recovery, but also signaled Turkey's desire to join the European Union.

- The mix of public and private company ownership is changing. Businesses are increasingly becoming well prepared for global competition, exemplified by the experience of Turkish Airlines.

- Foreign direct investment will increase substantially as more progress is made toward EU membership.

- Turkey's location is also extremely strategic with respect to the transportation of oil, not only for its domestic use, but also between the East and the West. Infrastructure is being developed to transport energy from the Caspian, Central Asian, and Middle Eastern locations to Europe and other parts of the world. For example, the Baku-Tblisi-Ceyhan pipeline was opened in the middle of 2006 to transport oil from the Caspian Sea. In 2007, the South Caucasus pipeline was in place to transport natural gas from Azerbaijan to Turkey. Also in 2007, a pipeline was implemented to transport Caspian natural gas to Europe via Turkey.

- The US and Turkey share common values and goals, as well as a solid strategic dialogue.

Air travel opportunities to, from, and within Turkey are enormous for some of the reasons cited above. Other reasons include: large segments of Turkish populations in many European states, (particularly Germany), almost 75 percent of the population living in Turkish urban areas (and the percentage is increasing), the expected performance of the economy that will undoubtedly raise living standards to those typical in Europe, and of course, the geographic location of Turkey. With respect to the last point, in some cases, the location of the Istanbul hub is even superior to the hubs under development in the Gulf region of the Middle East. This advantage can translate to shorter flight distances, for example, from Japan and South Korea. The advantage can also relate to the fact that, with the Istanbul hub, it is possible to use narrow-body equipment to fly passengers from Europe to the Indian sub-continent. Similarly, Turkish Airlines can develop its second hub, Ankara, whose geographic location can also provide the advantage of using standard, narrow-body aircraft to connect second-tier cities (like Birmingham, in the UK) to second-tier airports (like, Hyderabad, in India). Turkey also has a number of growing low-cost airlines, such as Anadolujet (a subsidiary of Turkish Airlines), Pegasus, and SunExpress (an affiliate of Turkish Airlines). The expansion of these airlines will lead to an increase in regional domestic services and tourism.

In light of the aforementioned reasons, and despite the fluctuations in the price of fuel, Turkish Airlines has been expanding its operations while most other airlines have been contracting. The airline's low cost structure has enabled it to maintain its growth strategy, even during the times of high fuel prices, and also to suffer much less with respect to the loss of premium traffic. The growth of the airline has also been helped by the growth of the country, its liberalization, the growth in tourism, and growth in the traffic connecting at its hub airport. It is the geographic location (55 countries located within 3.5 hours of flying time[17]) that provides an enormous potential for increasing the transit traffic, particularly, as already mentioned, when coupled with the fact that the flights can be operated with narrow-body aircraft and with high frequency. Turkish Airlines is a full-service carrier with low costs. With global presence, this could be an enormous competitive advantage.

The Paradox of Africa

Africa is perhaps the most controversial region with respect to challenges and opportunities for businesses. The following are just a few pessimistic perceptions plus a few optimistic perceptions.

Generalized Pessimistic Perceptions

- Although African countries achieved independence nearly half a century ago, the continent has failed to develop. It is still often viewed as a continent filled with drought, famine, poverty, disease, political unrest, conflict, and even war.

- It is the lack of principled, ethical leadership that has held Africa back, and continues to do so. It has been plagued with corruption, dictatorship, and instability, as well as war crimes, genocide, and crimes against humanity (such as in Darfur).

- Military rule continues to exist in some countries, such as after the military takeover of Mauritania in August of 2008 when the commander of the Presidential Guard overthrew the elected government. Moreover, even when a transition does come as a result of moving from military to civilian rule (as in the case of Nigeria), it does not seem to produce a change in ordinary people's lives.

- It is perceived that there has been more concern for outside interests than for its own people and those in power have not been connected or engaged with the vast majority of the African people. For example, consider those farming on the hillsides. There has been a failure to focus on any development that benefits them, such as education that would not only stop them from farming in ways that are not environmentally sound, but also educate them in ways to farm more sustainably.

- Diseases such as HIV/AIDS, malaria, and tuberculosis continue to claim too many lives, and are propagated by ignorance and misinformation.

- Even in South Africa, unemployment is reported to be high (23 percent).[18] There is also underemployment (employed people in households where each member lives on below US$1.25 per day). Such households make up almost 60 percent of the workforce in the sub-Saharan Africa.[19]

- There is a shortage of water. It has been reported that it can take up to 49 minutes per trip to obtain water in some areas of Africa.[20] Also relating to water, there are growing tensions over the Nile River, which is considered to be the economic lifeline for the region. Its waters are shared by ten countries in the east and north sections of the continent: Rwanda, Burundi, the Democratic Republic of Congo, Tanzania, Kenya, Uganda, Eritrea, Ethiopia, Sudan, and Egypt. In the case of Egypt, it is reported that it is almost completely dependent on the Nile, which provides more than 95 percent of its water requirements.[21]

- While some governments are addressing the perceived needs of its young people (and dabbling in social technology is a start), the real needs of the young include jobs, reduction in crime levels, and combating AIDS (which takes at least 1,000 lives each day in South Africa[22]).

- While there is oil in parts of Africa, the refineries are in poor condition (such as in Nigeria). This situation has led to insufficient oil for domestic use. Moreover, the economic benefit of oil exports has not affected or improved the lives of most people. Consequently, some oil-rich countries need to focus on how they can more efficiently and effectively use their oil and gas resources to better their own countries first, then to do so with profits attained through exporting.

- Many countries have let agriculture, one of their greatest assets, decline steadily through over use and poor policies. For example, instead of supporting and promoting their producers, some governments have put shackles on their producers. Contrast this situation with South Korea, a nation that was once poorer than Ghana, but has now caught up with the rest of the world with the aid of its government.

- Africa has not been able to use its foreign aid efficiently. It is reported, for example, that half of the world's aid had been reserved for Africa, even though India and China together have three times more people than Africa. Africa continues to remain uncompetitive compared with other developing nations. Consider, for example, that between 1970 and 1990, Africa lost half of its world market share to other developing countries, because those nations were able to produce and deliver the same goods more cheaply. It has been reported that this represented a loss of income for Africa of about US$70 billion per year.[23]

- Africa has an unfortunate shape. While it is the second largest continent after Asia, its coastline is barely one-quarter as long. It has few ports, which have been the key to economic development in other parts of the world.

- The infrastructure in Africa is very weak, and intra-regional transportation discourages internal trade among the countries within the continent.

Generalized Optimistic Perceptions

- The African Union (AU),[24] with its four pillars, has the potential to unite the continent through:

  1. *peace and security*—seeking to eliminate ongoing conflicts, prevent new ones, and prevent a relapse to conflict after peace is attained;

  2. *development, integration, and cooperation*—seeking a clear path to attaining middle-income status and a clear roadmap towards political, economic, social and cultural continental integration;

  3. *shared values*—seeking to promote shared values, including good governance, democracy, respect for human rights, accountability, and transparency;

  4. *institution and capacity building*—seeking synergy and collaboration, as well as a framework for sharing research and development initiatives.[25]

- A number of countries across Africa have experienced a peaceful change of power as the result of "normal" elections. Examples include Benin, Ghana, Guinea-Bissau, Kenya, Malawi, Senegal, Sierra Leone, and Zambia.

- There are examples of African countries which have experienced recent, increased social stability. For example, despite the terrible events that took place in Rwanda about 15 years ago, the country has not only achieved social stability, but has experienced significant improvements in income per capita and in the areas of education and healthcare.

- Libyan leader, Muammer Gaddafi, who has been in power for 40 years, has opened the country up to some of the world's largest oil groups.[26]

- There has been exponential growth in the mobile phone market; it has been reported that even in Africa, four in ten people now have a mobile phone.[27] This is also the case with respect to the implementation of fibre-optic cable in East Africa, a development that caused the region to go from having no undersea cable links to the rest of the world to now having three. On a broader level, the introduction of Information and Communication Technologies can facilitate the economic growth of a region, as evidenced by the experience of Nigeria.[28]

- Africa is of interest to India and China due to Africa's natural resources; China ran out of commodities in the late 1990s, and India will soon follow. Due to China's need for natural resources, it has been making enormous investments in the production of oil and gas from Angola to Algeria to Sudan, and is now importing more oil from Africa than from Saudi Arabia. As for exports, with consumer demand falling in the US and Europe, China is looking to Africa to be a buyer of Chinese manufactured goods. It is reported that, when one considers emerging market to emerging market trade in comparison to emerging market to advanced market trade, the emerging market to advanced economy share has declined from about 80 percent to about 50 percent, while the emerging market to emerging market share has doubled to around 50 percent.[29] Consider the following statistic: trade between China and Africa increased 45 percent from 2007 to 2008.[30]

- Africa possesses some highly desirable resources. Examples include their lucrative diamond fields (such as in Sierra Leone), oil (in Algeria, Angola, Libya, Nigeria, and Sudan), and natural gas (in Algeria, Egypt, Libya, and Nigeria). Zimbabwe is considered to have the world's second largest platinum, lithium, and corundum deposits, the potential to produce 30 tons of gold a year, and several high-grade diamond mines.[31]

- Africa also possesses an alternative to standard fuel. Specifically, two of Africa's poorest countries, Burkina Faso and Mozambique, are taking the lead as producers of the newly popular bio-fuel plant *jatropha curcas,* the seeds of which can be milled to produce an oil that works as diesel fuel. For example, around 62,000 farmers are now involved in a project in partnership with Germany's Deutsche Biodiesel.[32]

- Some countries, such as Morocco, have a strategic location with respect to both North America and Europe. Within Morocco, the city of Tangier is a natural gateway to Spain. Moreover, a number of other African countries, such as Tunisia, have relatively strong economic and commercial ties with Europe. For example, a Tunisia-EU program, launched in the mid-1990s, provided Tunisia with technical assistance and subsidies to help develop the Tunisian economy, leading to the elimination, in 2008, of customs tariffs and other trade barriers on manufactured goods.[33]

- Some countries, such as Morocco, which is considered to be a stable, democratic, and liberal Muslim state, are strategic players in terms of US interests in the Middle East.

- Nollywood of Nigeria, is now the third-largest film producer in the world, behind Hollywood and Bollywood. Over 2,000 films are made each year, and the industry is estimated to be worth more than US$250 million dollars.[34]

- Social progress is occurring in the cities such as in Nairobi, where once there was virtually no nightlife. Now the city boasts numerous bars and clubs.

- While still lacking in sufficient infrastructure, there is work toward a north-south corridor, in which Zambia's infrastructure and its cross-border connections will be the key focus. Similarly, Kenya's government has had discussions with China regarding the development of a port and transport corridor, which could provide a new avenue for Chinese oil in southern Sudan.[35]

- The continent is receiving attention from celebrities' aid efforts, such as those of actors George Clooney, Brad Pitt, and Angelina Jolie in Darfur, rock band U2's lead singer, Bono's efforts to increase trade and reduce poverty in Africa, and Microsoft's Bill Gates' efforts with respect to vaccinating children in Africa. In addition to these high-profile people, there are a growing number of "evangelical Christians, African entrepreneurs, Midwestern farm town families, ex-presidents and prime ministers" who are working hard to end the chronic hunger problem in Africa.[36] On the other hand, there are also some native African politicians and sociologists who claim that charity donations of all sorts can lead to corrupt regimes and hinder the development of self-sustainable economies.

Potential Business Opportunities

There is a whole other side to Africa that must be considered in terms of business opportunities. For example, consider the fact that Africa has more than 900 million consumers.[37] Obviously, the 900 million people belong to very different countries. The Arab part north of the Sahara desert has little in common with the sub-Sahara part. Moreover, in many countries there are three or four different tribes arguing about the influence over their countries (in some cases violently). Despite the enormous diversity within the continent, what would be some statistics, if one were to think of the continent of Africa as a single country?

- It would have had US$978 billion total gross national income (GNI) in 2006, thus placing it ahead of India.[38]

- Its GDP would show an average growth rate of 6 percent per year between 2001 and 2008.[39]

- It would rank as the tenth largest economy in the world, thus placing it ahead of three of the four initial BRIC countries (all except for China).

- It would have one of the youngest markets in the world, as more than half its population is under the age of 24.

At the heart of the opportunities that lie in Africa is "Africa Two," which is a group of approximately 400 million people. This is the rising middle class that is comparable to similar segments in India or China, and there is great interest in tapping into this future middle class. These people are aspiring to a better standard of living, and are upwardly mobile; they are the future of the African market. The higher standard of living will be facilitated by the increasing competitiveness of Africa that, in turn, will be facilitated by the opening up of markets across the region and through a more business-friendly environment. One key ingredient is the need for infrastructure that will be boosted by investments being made by China for the reasons cited in this chapter.

Successful companies have been able to recognize the opportunities present in Africa, if one is able to look beyond the above factors such as corruption, disease, and war. For example, Unilever increased its presence in Africa after tapping out opportunities in the US and Europe, where the company faced increasing competition as well as declining profits. Other companies, such as Coca-Cola, have been doing business in Africa, and have experienced the rewards, firsthand. Coca-Cola has had a presence in Africa since 1928. Despite fluctuations among individual countries in Africa, business has increased steadily over the past two decades. The company has enjoyed sales of 93 million beverages each day across the continent. In 2006, this translated to about US$4–5 billion in system revenues for the company, as well as its bottlers. However, in the case of Coca-Cola, the company has developed a strong relationship with stakeholders (including governments) to promote conditions for growth. Across Africa (from Egypt to Namibia with a concentration in East Africa), the Coca-Cola company has established 2,500 manual distribution centers that are independently-owned and employ 12,000 people. They provide opportunities for local entrepreneurs, function well within the local infrastructure, and provide enormous benefits for the local communities.[40] Similarly, IBM is making "a name for itself in the water market, and how its innovations may contribute to more efficient water management across the globe."[41]

However, it is not just major corporations that are finding opportunities in Africa, but also entrepreneurs. Asian companies represent one example. Specifically, the Chinese are offering a variety of products in Africa, everything from low-cost televisions and other appliances, such as generators, to clothing and shoes. Indian and Pakistani merchants also offer clothing, as well as leather goods. In some ways, this entrepreneurship could even be traced back to the Indian merchants who came into East Africa in the wake of British colonists. They played a similar role as the Chinese traders in Indonesia. One might ask why two members of the initial BRIC countries have looked to opportunities in Africa? One possible explanation is that China and India are able to recognize the potential in Africa, because they, themselves, have lived through a similar situation in a familiar marketplace. The Chinese and Indians have seen the rise of markets in their own countries, and therefore can see the same possibilities in Africa. In the case of China, the country looks at Africa as a major source of raw materials, and, as such, supports some governments in return for exploration rights.

With respect to entrepreneurship in Africa, it is important to note that, as business writer Vijay Mahajan has recognized, those who have found success have built their businesses around the economic, political, and military unrest. These entrepreneurs are not waiting for the situation to be organized and settled. Conducting business in such an environment requires great flexibility; entrepreneurs become problem solvers. For example, electricity is often unstable, so an entrepreneur may end up in the generator business. It is also important to note that reaching consumers in this market means offering the right products. For example, a large, expensive box of laundry detergent would not be appropriate for this market, as Procter & Gamble found out in Egypt, where the typical washing machine might be semiautomatic, made out of old barrels. The company needed to adjust their laundry detergent offerings accordingly. Furthermore, consider the fact that in Nigeria, as in other parts of sub-Saharan Africa, shampoo is sold through hair salons, and those who cannot afford the salons have their hair done on the streets. Most shampoo is not sold at retail outlets. This is a serious distribution challenge for traditional brands.

One particular area that is growing exponentially in Africa is the mobile phone market, as mentioned previously. Specifically, one study found that cell phones were achieving a compound annual growth rate of 85 percent or more in 10 of the 17 African nations surveyed.[42] One reason is the difficulty of providing and maintaining landlines. Another reason that cell phone growth has taken off at such an unbelievable rate is that it is a catalyst for the entire economy; other businesses are being built upon the cellular connections. For example, TradeNet, based in Ghana, created a free trading platform which involved several countries in West Africa. This platform allowed farmers to buy and sell their agricultural products. Companies are recognizing that cell phones present new opportunities for conducting business in such an environment. Specifically, a 2005 study found that the economic impact of cell phones in developing countries may be twice as large as in developed countries. This can be contributed to the fact that since there are few landlines, cell phones are the first communication infrastructure in many areas, providing a foundation for small businesses and connecting rural areas to the world.[43]

Finally, entrepreneurship in Africa could accelerate due to another recently observed trend, namely, a shift in global migration patterns. Some countries (such as Angola, Ghana, Kenya, and Nigeria) have begun to see a reversal of "brain drain." Some of the educated professionals who left their countries to work in developed countries appear to be returning to their native lands. These returning professionals could facilitate a growth in businesses, small and large.[44]

Consequently, if focused entrepreneurs can adapt around the constraints, the landscape would change when governments begin to change regulations to facilitate the development of business. In the World Bank's report, *Doing Business 2010*, a sub-Saharan country, Rwanda, was listed as the fastest reforming country. Its rank moved from 143 to 67 from 2009 to 2010.[45]

Challenges and Opportunities for Airlines

In some ways, Africa is ideally suited for the expansion of the air transport industry. First, there is a huge population; approximately 900 million. Even if only 10 percent of this population were to be able to take advantage of some truly basic air transportation services, the market segment would be greater than the population of the largest country in Europe, and about one-third of the population of the US. Yet, despite this huge population, Africa accounts for less than 5 percent of the total global air travel. Putting aside the size of the population, about one-third of the countries within Africa are land-locked. This geographic feature of the continent provides ample opportunities for the further development of air transportation. Next, air transport is in some cases the only viable mode for consumer goods to be available in international markets, especially agricultural products. If the world is globalizing, then air transport will enable Africa to participate in this movement. Finally, development of the air transport sector can facilitate the integration of Africa and promote the vision of the African Union.

Unfortunately, the development of the air transport sector has been facing numerous challenges for decades. These challenges lie in the areas of thin markets, insufficient liberal regulatory policies, lack of capital in the hands of airlines (other than the top three or four airlines), safety, air traffic management, navigation aids, and security. Technical assistance, as well as financial resources and management know–how, have been provided by organizations from the developed nations, but they have been provided on a piece-meal basis. This piece-meal approach is, in part, the reason for the continuation of the challenges for such a long time. From time to time, comprehensive approaches for solving the problems facing commercial and operational aspects of aviation have been suggested, but progress has been extremely slow. For example, the aviation community in Africa did adopt the multilateral Yamoussoukro Decision[46] to liberalize the air transport sector, remove barriers to the growth of the air transport sector, and improve the infrastructure and services in Africa. However, the pace of progress has been extremely slow, placing the continent well behind other developing regions. Outside assistance can help, but it will not resolve Africa's challenges without increased desire and cooperation from the African people themselves.

Briefly, the availability of scheduled services is poor and the costs of travel are higher. Basically, there are four major airlines: Egyptair, based in Cairo in the north; South African Airways, based in Johannesburg in the south; Ethiopian Airlines, based in Addis Ababa in the east; and Air Maroc, based in Casablanca in the west. The next six airlines are, alphabetically, Air Algérie, Air Mauritius, Atlas Blue, Kenya Airways, Libyan Airways, and Tunisair. Except for the top-tier carriers, such as South African Airways and Egyptair, these carriers are undercapitalized and have disproportionately higher operating costs. Moreover, except for air travel out of the top dozen or so airports linked to international destinations, the cost of travel within the continent tends to be high, and services are relatively poor. For example, it is not uncommon for a person living in the eastern part of Africa to fly north to Europe to get to a West African destination. This problem exists due mainly to the low density of passenger traffic within the continent. The top ten airports are Johannesburg, Cairo, Cape Town, Sharm el-Sheikh, Casablanca, Hurghada, Addis Ababa, Algiers, Lagos, and Nairobi. Figure 6.4 shows the location and the size of these airports (along with nine other airports) relative to the size of the largest passenger-handling airport in 2008, namely, Johannesburg. The airlines and airports at the low end of the scale are reported to have poor facilities and services, such as the lack of electronic tickets and machine-reading passport devices.

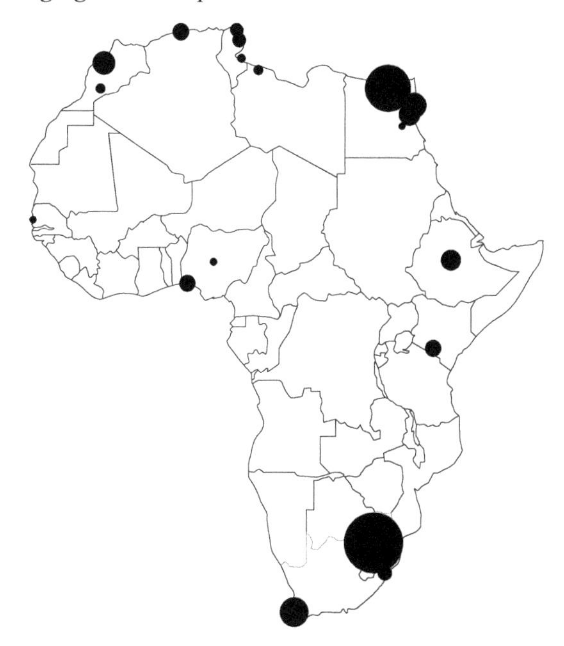

Figure 6.4 The Top 19 Passenger-Handling Airports in Africa, 2008

As mentioned earlier, neither the challenges nor the opportunities facing the aviation industry in Africa are new. They have been discussed for many years. For example, there was a Seminar for African Airline Chief Executives arranged by the IATA/AFRAA held in Nairobi in September 1984. In this seminar, there was significant discussion on the commercial and technical problems facing the management of African airlines and the concept of cooperation as a means to solving such problems.[47] However, one reason the challenges have occurred for so long is that the proposed approach has been piece-meal as opposed to comprehensive. However, various stakeholders are now more likely to identify and implement much more comprehensive approaches for the following reasons.

- Other emerging markets are progressing at a much faster rate than Africa in developing their air transport sector, as exemplified by the developments in Brazil, China, and India.

- The movement of goods by road is not only slow due to poor infrastructure, but also expensive due to high freight rates relative to other developing regions. The development and maintenance of an integrated pan-African railway network will take a much longer time and cost more than a far more efficient air transport system. Keep in mind that in some cases the railroads built in the past 40 years have been deteriorating, exemplified by the experience of TaZaRa (Tanzania-Zambia Railroad) and the railroad between Addis Ababa and the port of Djibouti. Consequently, there are enormous opportunities for growth in the cargo sector, evidenced by the experience of Ethiopian Airlines.

- Africa is beginning to recognize that globalization is proliferating at a much higher pace and one way for Africa to participate in this phenomenon is to boost its commercial aviation industry.

- African governments are beginning to believe that air transportation is an indispensible engine and a catalyst to promote their economic growth strategies and initiatives.

- Governments are beginning to see that the resistance to open up skies in regional markets, coupled with the reluctance to cooperate, has been reducing even further the share of the global market transported by African carriers.

- Some countries are recognizing that if the US continues to push the Open Skies framework and the EU wants to negotiate as a block, then the AU must move much further in defining a common external air transport policy for Africa. A framework for cooperation among airlines within Africa must be created to develop intra-African services, improve safety, and reduce costs.

- The African carriers are becoming even weaker, given the progress of the consolidation process in other regions (mergers and acquisitions, as well as further integration within alliances).

- The 2008–2009 economic downturn, as well as the increase in competition in other regions of the world, is encouraging carriers from the developed regions to transfer their excess capacity to destinations in Africa. Consequently, Africa-based carriers need to strengthen their positions to pre-empt such initiatives.

- It is clear from the overall experience in South Africa (leaving aside the problems faced by South African Airways during the 2007–2009 period) that an economically viable, commercial, air transportation system can be developed to not only improve integration within the country, but also to participate in the globalization process.

- An increase in the size of operations of established airlines in Africa, such as Egyptair and Ethiopian, will lead to a reduction in operating costs, as well as a diversification of risk during economic slowdowns.

- Pre-owned small aircraft (for example, the 50-seat regional jets) can be quite suitable for many thin markets. Such markets could be domestic, such as between Kaduna and Port Hartcourt in Nigeria, and between Laayoune and Rabat in Morocco, or they could be between African countries, such as between Abidjan, Côte d'Ivoire and Nouakchott, Mauritania, between Durban, South Africa and Harare, Zimbabwe, and between Accra, Ghana and Abuja, Nigeria.

- Countries are beginning to deregulate their domestic markets (for example, Egypt, Morocco, and Nigeria), enabling the creation of new airlines, such as Air CEMAC, ASKY, Virgin Nigeria, and Star Equatorial, to start offering more services within the region.

- The Cape Town Convention and Aircraft Protocol (concluded in 2001) offer hope for facilitating the modernization of airline fleets. The convention provides potential aircraft financiers with some key rights that, in turn, provide an incentive for financiers to offer lower financing.

- Emerging technology will have a much greater impact on airlines operating in Africa relative to those operating in advanced economies where the airlines have been benefiting from technology on a gradual basis. The airlines in Africa will end up leapfrogging into the 21st century, and, by doing so will achieve a step-function benefit.

For some of the aforementioned reasons, it is possible to imagine that one or two of the top airlines in Africa (with the financial resources and the management and technical know-how) could take the initiatives to set up satellite airlines within Africa at strategic locations, following the example of LAN Chile. Examples of such locations include Luanda in Angola, Libreville in Gabon, Accra in Ghana, Dzerba in Tunisia, Benghazi in Libya, and Entebbe in Uganda. Satellite airlines could be new airlines based in these countries, or small existing airlines that could be expanded. It is assumed that in each case, the appropriate government will step up the pace of structural reforms to facilitate the development of such an initiative. The progressive governments are likely to implement necessary changes to their national laws to facilitate greater cooperation and bring greater efficiencies to the air transport sector. Cooperation is needed for a variety of reasons and in a number of areas. Many markets are very small, making the economic viability much more of a challenge in terms of fleet and network expansions that, in turn, relate to bilateral and multilateral agreements. Cooperation is also needed to develop adequate control, navigation, surveillance, and air traffic management (CNS/ATM) systems to provide safe and smooth air traffic flow over much of the space of Africa. Next, cooperation is needed in the area of private/public partnerships in the development of the air transport sector. Finally, cooperation is needed to reduce tariffs on the continent.

Is there any evidence that such a scenario is possible? The answer is a clear "yes." A new airline, ASKY, was conceptualized in 2008 in Ouagadougou, Burkina Faso that would have the potential to foster African integration beginning with the implementation of daily flights to all sub-Saharan countries. Subsequently, the decision was made to base the airline in Togo's capital, Lome, with Ethiopian Airlines taking 25 percent equity and, if necessary, providing the aircraft to begin operations. It was also reported that Ethiopian Airlines would develop a hub in the region of Lome to connect passengers to other countries using narrow-body aircraft, such as the Boeing 737.[48] Eventually, it is foreseen that ASKY would expand its operations to longer-haul, international markets using aircraft such as the Boeing 767. The key point to notice is that ASKY's model for West Africa (a sub-region that accounts for almost one-third of the total population of Africa) could be duplicated and based in other parts of Africa, such as in the cities mentioned above. Just as LAN Chile played a key role in the development of a unique, strategic alliance in Latin America (see the discussions in Chapter 4 and later in this chapter), Ethiopian Airlines could achieve similar results in Africa.

Next, is there any evidence that the air transportation system is improving, and that there are real opportunities for airlines? Consider the actual and potential developments in Africa. The economic landscape is changing in Nigeria. Nigeria has a population of about 140 million, and it is now the eighth largest oil producing country in the world. These two factors alone are sufficient reasons to expect huge opportunities in the growth of the air transport sector. Despite the poor infrastructure hindering the full potential of the oil industry, and despite the continued existence of high levels of corruption, the oil and non-oil sectors of the industry have been fuelling the growth of the economy. Gross Domestic Product increased in 2008, income per capita increased by 25 percent, foreign reserves grew, and the food supply has remained stable.[49]

As a result of the growing economy, related almost wholly to the oil sector, there has been significant growth in air travel and the airport in Lagos has been developed. On the airline side, Lagos, with a population that is approaching 20 million, is served by such international carriers as Air France-KLM, Alitalia, British Airways, Delta Airlines, Lufthansa, and Turkish Airlines. These carriers are transporting high-yield business traffic. On the domestic side, it is reported that the growing middle class is generating a significant amount of traffic. On the airport side, the new Lagos terminal, developed through the first public-private partnership project in Nigeria, is reported to have set new standards for customer service and satisfaction (air conditioning, restaurants, shops, ATM machines, a planned Intercontinental hotel) and operational efficiency using state-of-the-art technology (such as for facilitating check-in). It is interesting to note that 85 percent of the passengers traveling through the domestic terminal are Nigerians, and only 15 percent are foreigners (mostly Europeans, Chinese, Indians, and South Africans).[50] While, it is true that the media is reporting continuous problems relating to crime and chaotic infrastructure, reports are also emerging on other aspects that signal the potential of Lagos to become a megacity. For example, Babatunde Fashola, the Governor of Lagos State, is reported to be on a mission "to transform Nigeria's dysfunctional and chaotic commercial capital into Africa's model metropolis."[51] Assume that he is able to achieve his mission with the help of investments from Nigeria's rapidly-growing banks, and others, to dramatically improve the infrastructure, and by persuading people to support him to promote civic pride. Assume also that the UN's projected population of Lagos becomes a reality making it the third largest city in the world by 2015 (after Tokyo and Mumbai).[52] Under this scenario, one can only imagine the growth possibilities for the air transport sector.

These possibilities are amply substantiated by the achievement of two rising stars, Ethiopian Airlines and Kenya Airways. Both airlines are capitalizing on their geographic location and the opportunities provided by the globalization process to expand their operations. Ethiopian Airlines, for example, serves almost three dozen cities within Africa, half a dozen cities in Europe, more than two dozen cities in the Gulf, Middle East, and Asia, and even one in the US. The carrier has placed orders for modern aircraft and has generated net profits for the fiscal year ending March 2009.

The Paradox of Latin America

Much of the US population views Latin America, a region with a population of more than 500 million (of which a high percentage are urban dwellers) in terms of certain images, such as dictators, drug lords, and illegal immigration. These generalizations, based on such images, are the result of ignorance. There is another face of the region, a resurgence as a result of civilian elected governments and their more market-oriented policies, including the policies to accelerate global integration. As in the case of Africa, this section provides a synopsis of both viewpoints as well as the potential challenges and opportunities for airlines.[53]

Generalized Pessimistic Perceptions

- Guerrillas, drug cartels, violence, lawlessness, military dictatorships with personal agendas and policies, authoritarian leaders paying only lip service to democratic ideology, border clashes, illegal immigration, high levels of poverty, extreme levels of income inequality (even higher than in Africa) in some regions, (as in the case of Africa) an extensive illegal commodities trade which, in turn, threatens any real long-term economic progress, and nightmarish bureaucracy.

- Lack of futuristic vision and planning from some leaders has resulted in poor and dated infrastructure, constraining personal well-being as well as prosperous business development.

- An entirely different cultural mindset with respect to work, leisure, concept of time, and business practices (for example, searching for extreme, quick solutions) than those of North America or northern Europe, plus a long history of a culture dominated by males.

- Widespread corruption/bribery at high levels (political leaders and their aides) and at petty levels (police officers and traffic violations) have not enabled ordinary people, especially the indigenous segments of the populations, to benefit by raising their standards of living, even in regions that are rich in resources; hyperinflation, currency devaluation, and debt defaults. High levels of bureaucracy and corruption also make it difficult to start up and grow sound businesses.

- Inability to achieve significant integration, despite some favorable attributes, such as only two major languages, territorial proximity, and no major military tensions, making it (as in the case of Africa) economically uncompetitive with low productivity, low innovation, low education standards, and high business costs.

- Many countries are still trying to define their socio-political system (liberal, socialist, center-right), leading to uncertainty, in turn, related to rapid-changing and hard-to-forecast market conditions for conducting business. This lack of political and judicial stability has kept many foreign investors and business travelers away.

Generalized Optimistic Perceptions

- Latin America is one of the richest continents with respect to natural resources of all kinds. This advantage will grow in importance as the world becomes more resource-constrained. Added to this inherent advantage is another fact, namely, its natural settings for tourism that have been underexploited thus far.

- There were successive financial crises starting with the 1980s (which has been referred to as "The Lost Decade") when the region was plagued with high levels of debt and hyperinflation. However, beginning in the 1990s, there has been an increased presence of civilian elected governments, increasing accountability and transparency in government. The presence of female leaders has also increased, representing a fundamental shift from a region long ruled exclusively by men. Michelle Bachelet took office in 2006 as President of Chile, and Cristina Fernandez de Kirchner took office in 2007 as President of Argentina.

- Although improvements have been slow in coming relative to other emerging markets, such as China and India, the potential is gaining momentum. The region is beginning to install new governments committed to the implementation of economic, financial, and legal reforms. These initiatives will promote growth and make economies resilient through privatization of public enterprises, transparency, accountability with strengthened checks and balances, market-determined foreign exchange rates, judicial independence and upgraded bankruptcy laws, more autonomous central banks, and freedom-of-information laws kept in line via investigative journalists.

- Inflation has stabilized in the single digits, partly due to the public's recognition of the need to control inflation and partly due to targeted monetary policies.

- Latin America has a substantial level of proven oil reserves after the Middle East, with Venezuela being the top producer (similar to Nigeria in Africa). Other oil exporting countries include Brazil, Ecuador, and Mexico. Although some of these countries may not have taken full advantage of the high price of oil during 2008 by increasing their output, there is a potential to increase the efficiency of operations, and attract investments to improve the infrastructure. In addition to oil, there are many countries, such as Mexico and Peru, that are rich in commodities.

- Latin America does not have the serious problems facing Africa (famine, genocide, AIDS, etcetera). In addition, the region is not associated with nuclear concerns (such as Iran and North Korea) or Islamic threats.

- Although, in the past, there was not much interest in Latin America by global countries (except by the US), now there is increasing interest by countries, such as China and India. As in the case of Africa, China has a strong interest in commodities, such as iron ore and soybeans, and is willing to pay high prices for them.

- As for corruption, the region is finding ways to overcome it to compete more effectively in the globalizing economy. Consider the fact that, while corruption also exists both in

China and India, the economies of these countries have been growing with the aid of government policies and through the education of its populations. Chile and Uruguay, for example, have been ranked among the 50 "cleanest" countries worldwide.[54]

- Strong relations with Latin America could facilitate the self-sufficiency of energy in the Western Hemisphere and some independence from the turbulence of the Middle East. For example, oil from Canada, Mexico, and Venezuela, coupled with new sources of power (such as Brazilian ethanol), could unite North and South America in a powerful trading bloc.

- Latin America has the potential to become a powerful economic region, specifically with the resources of Brazil and Mexico. Moreover, in the case of Mexico, from a longer-term perspective, its proximity to the US is a major plus with respect to a market for Mexican exports. In addition, to their large populations, Brazil and Mexico have the highest GDP/Capita (adjusted for purchasing power parity) within Latin America.

- There has been a growing impact of Latin American names in all walks of life worldwide (artists, business executives, sports people, etcetera), especially in the US (which resonates globally).

Potential Business Opportunities

In 2007, Antoine van Agtmael wrote a book, *The Emerging Markets Century: How a New Breed of World-Class Companies is Overtaking the World,* in which he listed 25 of the top emerging multinational giants. Fourteen of these 25 companies are based in Asia, ten in Latin America, and one in Africa. Out of the ten in Latin America, four are in Brazil (Aracruz, Embraer, CVRD, and Petrobras), four in Mexico (Cemex, Modelo, Televisa, and Telmex/America Movil), one in Chile (Concha y Toro), and one in Argentina (Tenaris).[55] Based on the success of these businesses, it is clear that, despite the challenges listed in Latin America, there are also enormous business opportunities for companies who not only are visionaries of the future of globalization, but who are also willing to develop innovative strategies around the constraints that exist in Latin America. Examples of constraints are poor distribution channels, erratic regulations, and corruption. The important point is not only that there are enormous business opportunities for companies in the developed regions as evidenced by US-based GM and Whirlpool, but also that if the companies in the developed region are not careful, companies based in the emerging markets could capitalize on these market opportunities, pre-empting the companies based in the developed regions.

To appreciate the market opportunities in Latin America, one needs to look at only a few countries. Consider Panama. Its economy is dollar-based, has low inflation, has grown at a stable rate, and is diversified. The initiative to expand the Panama Canal has created not only a huge number of jobs for many years to come, but it will also ensure a huge revenue flow in the future. The connectivity (rail, ocean, air) feature, coupled with the geographic location of Panama, has the potential of making Panama the "Singapore of Latin America." The favorable government policies and the ease of conducting business are enabling Panama to become a major international business center, as evidenced by the growth of the banking industry, and the establishment of regional offices by international companies, such as P&G and Caterpillar. In addition to being on the way to becoming a business center, Panama is also becoming an important tourist center. Panama has a GDP/Capita (adjusted for purchasing power parity) of over US$11,000 and it has the highest propensity to travel among the major countries in Central and South America.[56]

Similarly, while Mexico is experiencing a significant downturn in the current recession, it has enormous resources that will help it to bounce back quickly. Besides having a population that exceeds 100 million and oil resources, it has developed an enormous production capability (that had been used by the US auto industry, for example), a relatively low labor wage rate for the well-educated workforce, and a close proximity to the US, not to mention a partnership in NAFTA.

Then there is Brazil, a country with a population of 200 million, an enormous, recent reduction in poverty, and a well-developed economy in segments such as agriculture, mining, manufacturing, and services. With respect to the segment of the population that is still considered impoverished, the social program, "Bolsa Família," is reported to have raised 24 million people above the poverty line. Just as important, the Brazilian middle class is definitely growing, partly from the improvements in the economy, and partly because of the expansion of social programs. The Brazilian government's social initiative, "My House, My Life," provides cash grants and discounted mortgage rates toward the purchase of homes.[57] The growth of the middle class is enabling existing companies to expand, and new businesses to be started. Grupo Pão de Açúcar (a chain of almost 600 supermarkets with sales approaching US$10 billion) is one example in the first category. The airline, Azul, is an example in the second category.[58]

While Latin America cannot be viewed as a single market of more than 500 million people, and while there are some individual countries with leaders with their own personal agendas, the region, as a whole, has enormous market opportunities provided by the countries with democratic governance and who have embraced globalization through an outward orientation. Brazil, Chile, and Colombia are representative examples discussed in this section. Moreover, it is important to keep in mind that Latin American consumers are eager to consume premium products if offered at reasonable prices. People are purchasing Blackberries in Chile, Chivas 18 in Venezuela, BMWs in Panama, and designer clothes in Colombia.

Challenges and Opportunities for Airlines

Major challenges facing the air transport sector in Latin America relate to the small size of the markets, tight government control of market access, as well as ownership and control, inefficiencies of the aviation infrastructure (airports and the air traffic control systems), as well as, in some cases, the commercial and operational aspects of airlines themselves. Take, first, the case of thin markets. In Central America, the markets were so thin that no local airline in the region could operate an economically viable network. This challenge led to the development of alliances unique to Latin America. Grupo TACA (Transportes Aéreos Centroamericanos—Central American Air Transport) was formed in the 1990s by combining the tiny airlines from each country in Central America. The group began to operate as a single brand focused on the transportation of passengers only. Economies of scale were achieved through the combined purchase of an all-Airbus 319/320 fleet. Similarly, the only viable way for LAN Chile to grow within a restrictive and protectionist regulatory environment was to acquire stakes in airlines based in neighboring countries, to develop a common brand, and to implement strategies to achieve economies of scale. See LAN's example in the last chapter relating to its integrated network. LAN (Latin American Network) is now one of the most successful airline business models in a region with thin markets and tight government controls.

With respect to government control, the issue is not just related to restrictive and protectionist regulatory policies, but also to the uncertainty of the political situation. Here is one example. The liberalization phenomenon that started in the US in the late 1970s, and continued in Europe, ultimately began to have some impact on the air transport sector in Latin America. For example, the Brazilian government initiated a liberalization process in the 1990s. The liberalization movement increased competition, not only among the existing network carriers, but also brought in the domestic marketplace in 2001, GOL, a low fare airline. The older main network airlines in Brazil could not survive the increase in competition. Transbrasil ceased operations in December 2001. Partly as a result of this event, in 2003, the Brazilian government decided to re-regulate its airline industry to control an alleged excess capacity, that, in turn, was the result of excessive competition. Still, VASP ceased services in January 2005, and Varig was purchased by GOL in 2007. Accompanying the demise of these three traditional network airlines, another low fare airline, Azul, entered the domestic marketplace in 2008.

Another major challenge has been for the government to develop sufficient infrastructure in Brazil to keep up with the increase in demand. Consider one example, the situation in São Paulo, which has the two busiest airports in Brazil, Guarulhos

(GRU) and Congonhas (CGH). There is insufficient capacity at the airports and within the air traffic control system. The two airports are separated by about 20 km, and the terminal area has experienced significant congestion, especially since the two airports have converging runway patterns.[59] Similarly, there is insufficient capacity at the two airports in Rio de Janeiro, at the Galeão (Antonio Carlos Jobim International Airport) and at Santos Dumont Airport. The challenge is either to increase the capacity of the airports (new airports, additional runways at existing facilities and investment in new technologies), or to better manage demand (for example, better management of air traffic flow and the allocation of slots).

The infrastructural issues continue to exist in many parts of the continent. For example, on the airport side, not only is there a need for the modernization of airport facilities, but the level of fees is relatively high, and remain high even when a government privatizes an airport. One reason is that some governments still collect funds from the private companies and use these funds for non-aviation purposes, instead of re-investing them to improve the aviation infrastructure. Another daunting obstacle has been related to safety. Some Latin American airlines were barred from operating to the US under the International Aviation Safety Assessment (IASA) program. According to the US Federal Aviation Administration, these airlines did not comply with the safety oversight standards established by the International Civil Aviation Organization.

On the opportunity side, one can start with the size of the population of the region. Even if one assumes that only 10 percent of the population can afford to fly, this segment represents about 50 million people. Consider Brazil as an example. There is an enormous demand for airline service from the non-business segment if services can be provided at reasonable fares. According to one estimate, 50 percent of the Brazilian population can be considered to be in the middle class.[60] According to another estimate, there are more than 150 million people who currently travel long distances by bus to save money. Let us not overlook the opportunities within the business segment if services could be provided in less dense markets on a nonstop basis, and with reasonable frequency.[61] In other words, connect the dots directly.

Looking at Latin America as a whole, there are probably 2000 markets that could be served to meet the needs of the marketplace. Table 6.1 lists the top 20 markets by the number of airline seats offered in April 2007 and the number of seats added in April 2008. As evidenced from the data shown, most of the largest markets are either within Brazil and Mexico, or from other points in Latin America to Brazil and Mexico. Consequently, there are enormous opportunities for further growth within the air transport sector to meet the needs of the 200 or so million people that live in countries other than Brazil and Mexico. One of the best examples would be Chile. Not only does Chile have the opportunity to showcase and grow its tourist industry, but the business travel to and from Chile is likely to grow at enormous rates, as it is becoming South America's new IT hub. Read, for example, the article in the August 2009 issue of *BusinessWeek*: "U.S. tech companies are flocking to Chile, with its savvy engineers, low wages, and government grants."[62] A similar case can be made for Panama. It is interesting to note that the Latin American air travel market is still growing, and the national carriers, LAN and COPA, made profits even during the downturn in 2008. Likewise, from the viewpoint of tourism, Colombia has literally re-branded itself. One only needs to see the increase in the number of cruise ships including a stop at the Spanish colonial port town of Cartagena.

However, Latin American carriers should try to expand their own operations to get closer to an equal share of the market. Consider, for example, the following statistics. In 2008, North American airlines transported 84 percent of the passengers between North America and Latin America and the Caribbean, compared to the 16 percent transported by the Latin American and Caribbean airlines. This presents substantial opportunities for Latin American carriers to serve un-served or under-served markets in the US. Examples include from Boston in the US, to both Bogota and Panama. Similarly, the European airlines transported 79 percent of the passengers between Europe and Latin America and the Caribbean, compared to the 21 percent carried by Latin American and Caribbean carriers.[63] As governments transform their policies to improve this imbalance, enormous opportunities for Latin American airlines will be created. However, while there are opportunities to balance these market shares, Latin American carriers can also not overlook the needs of intra-regional markets, particularly those in the category of the second tier.

Table 6.1 Top City Pairs in Latin America: April 2008 Comparisons vs. April 2007

City Pair	Total Seats	Seats Added
Rio de Janeiro-São Paulo	360,708	18,861
Mexico City-Monterrey	171,483	58,387
Guadalajara-Mexico City	146,153	42,894
Brasilia-São Paulo	142,982	21,204
Cancun-Mexico City	129,629	38,124
Belo Horizonte-São Paulo	129,510	39,998
Porto Alegre-São Paulo	124,122	14,942
Curitiba-São Paulo	120,618	3,934
Brasilia-Rio de Janeiro	98,998	37,272
Salvador-São Paulo	98,570	41,095
Bogota-Medellin	95,278	2,928
Guayaquil-Quito	92,284	-10,620
New York-San Juan (PR)	86,704	-9,230
Bogota-Cali	85,795	-1,866
Florianopolis-São Paulo	79,382	16,390
Buenos Aires-São Paulo	72,779	13,820
Buenos Aires-Santiago (CL)	72,482	-13,399
Belo Horizonte-Rio de Janeiro	66,750	29,068
Merida (MX)-Mexico City	62,204	10,952
Rio de Janeiro-Salvador	58,260	14,396
Remaining (1898) average	6,207	884

Source: Prepared by ALTA with information from the OAG and printed in ALTA's Anuario 2008 Yearbook, p. 18.

Takeaways

Despite the daunting obstacles that have existed and continue to exist in both Africa and Latin America, the air transportation sector has a bright future on both continents. While it is true that further development of air transportation system is dependent upon the achievement of political, social, and economic stability, it is also true that the development of air transportation will, in turn, facilitate the social and economic development. The impetus is beginning to come from trend-setting airlines, such as Ethiopian and Kenya Airways in Africa and COPA and LAN in Latin America, with bold and innovative initiatives. However, the airlines based in both regions should not replicate the past mistakes of other airlines based in other parts of the world. This is their opportunity to tailor air transportation to their marketplace realities. Some airlines need to have business models that are appropriate for their lower-income populations.

The financiers can help airlines in the areas of the rationalization of their shape and size. Both Africa and Latin America will see an intensive consolidation process that could lead to fewer, but stronger, airlines. The actions of the financiers will facilitate the consolidation process. They are not likely to provide capital to weak carriers and, even in a few cases where capital does become available, the cost of such capital would be very high. Once the weaker carriers are out of the marketplace, the surviving carriers will become stronger resulting in even more access to capital becoming available and at a lower cost. Finally, progressive governments will recognize the catalytic role of the commercial aviation industry to facilitate economic development. As evident from this chapter, airlines can expect numerous "pockets of growth" around the world, centered on metropolitan areas, or strategically-located hubs, such as those in the Gulf region of the Middle East. Open skies regimes coupled with advanced-technology airplanes (discussed in the next chapter) will enable airlines to meet the needs of these population groups with direct airline service.

Notes

1 World Economic Forum, *The Global Competitiveness Report*, 2008–9, p. 9.
2 Harteveldt, Henry, Forrester Research, Presentation at the OSU International Airline Conference, Las Vegas, April 2009.
3 Mahajan, Vijay, *Africa Rising: How 900 Million African Consumers Offer More than You Think* (Upper Saddle River, NJ: Wharton School Publishing, 2009), p. 38.
4 Marber, Peter, *Seeing the Elephant: Understanding Globalization from Trunk to Tail* (Hoboken, NJ: John Wiley, 2009), p. 9.
5 Aglionby, John, "Economists Split on Jakarta's Komodo Power," *Financial Times*, September 10, 2009, p. 5.
6 World Economic Forum, *The Global Competitiveness Report*, 2008–9.
7 http://www.state.gov/r/pa/ei/bgn/1981.htm, pp. 1 and 2.
8 http://www.state.gov/r/pa/ei/bgn/1981.htm, p. 6.
9 To Chile, with its savvy engineers, low wages, and government grants," *BusinessWeek*, August 24 & 31, 2009, p. 73.
10 Material for this section was derived from the following sources: Goodwin, Paul B., *Latin America* (New York, NY: McGraw-Hill, 2009); Tuller, Lawrence W., *An American's Guide to Doing Business in Latin America* (Avon, MA: Adams Business, 2008); https://www.cia.gov/library/publications/the-world-factbook; http://www.state.gov/r/pa/ei/bgn/1981.htm; World Economic Forum, *The Global Competitiveness Report*, 2008–9.
11 World Economic Forum, op. cit.
12 Material for this section was derived from the following sources: Khanna, Parag, *The Second World* (NY: Random House, 2008); https://www.cia.gov/library/publications/the-world-factbook; http://www.state.gov/r/pa/ei/bgn/5309.htm; World Economic Forum, op. cit.
13 World Economic Forum, op. cit.
14 Material for this section was derived from the following sources: Khanna, Parag, *The Second World* (NY: Random House, 2008); https://www.cia.gov/library/publications/the-world-factbook; http://www.malaysia.gov.my/EN/Relevant%20Topics/IndustryInMalaysia/Business/ICT/MSC/Pages/MSC.aspx; http://www.state.gov/r/pa/ei/bgn/2777.htm; World Economic Forum, *The Global Competitiveness Report*, 2008–9.
15 The World Bank Group, *Doing Business 2010*, p. 4.

16 Ballantyne, Tom, "I Saw a Crisis Coming," An Executive Interview with Datuk Seri Idris Jala, Managing Director of Malaysia Airlines, *Orient Aviation*, December 2008–January 2009, pp. 24–7.

17 Sobie, Brendan, "What Slowdown?," An Interview with Dr. Temel Kotil, CEO of Turkish Airlines, *Airline Business*, June 2008, p. 34.

18 Source: South African government report, as reported in an article, "Time of Reckoning," by Mark Gevisser in *Focus on Africa*, April-June 2009, p. 18.

19 *The Africa Report*, June-July 2009, p. 12.

20 Ibid., p. 11.

21 Ibid., p. 38.

22 Lakaje, Mpho, "X Generation," *Focus on Africa*, April-June 2009, p. 21.

23 Calderisi, Robert, *The Trouble with Africa* (New York, NY: Palgrave MacMillan, 2006), p. 18.

24 The Organization of African Unity (OAU) was organized in the 1960s. The African Union (representing 53 states) was established in 2002 as the successor to OAU. The AU's secretariat, the African Union Commission, is based in Addis Ababa, Ethiopia.

25 *The Africa Report*, June-July 2009, p. 33.

26 "Triumphal in Tripoli," *Financial Times*, September 1, 2009, p. 5.

27 "The Power of Mobile Money," *The Economist*, September 26, 2009, p. 13.

28 "ICT: The Critical Success Factor in Creating Competitive Advantage Through Knowledge," in *This is Africa: A Global Perspective*, June 2009, pp. 14–5.

29 "Beyond the Silk Route," a Special Report in *This is Africa: A Global Perspective*, FT Business, July 2009, p. 7.

30 Abkowitz, Alyssa, "China Buys the World," A Special Report on China in *Fortune*, October 26, 2009, p. 95.

31 Norbrook, Nicholas, and Frank Chikowore, "Bring in the Earth-Movers," *The Africa Report*, No. 18, August-September 2009, p. 42.

32 Misser, François, "Jatropha: Oil from the soil,"*The Africa Report*, June-July 2009, pp. 93.

33 http://www.state.gov/r/pa/ei/bgn/5439.htm

34 "Widescreen Appeal," *Focus on Africa*, April-June 2009, pp. 58.

35 Jopson, Barney, "Kenya and Beijing in Talks to Build Port and Transport Corridor," *Financial Times*, October 15, 2009, p. 1.

36 Thurow, Roger and Scott Kilman, *Enough: Why the World's Poorest Starve in an Age of Plenty* (NY: Perseus Books, 2009), front and back jacket flaps.

37 Mahajan, op. cit., p. 3.
38 According to World Bank data.
39 Schwab, Klaus, "Improving African Competitiveness," This is a Special Report in *This is Africa: A Global Perspective*, FT Business, July 2009, p. 3.
40 "A Commitment to Development," a Special Report in *This is Africa: A Global Perspective*, FT Business, July 2009, pp. 12–3.
41 Nunes, Sharon, Vice President of Big Green Innovations at IBM, in "Using Technology to Transform the Water Market," in *This is Africa: A Global Perspective*, June 2009, p. 57.
42 A 2007 study by the Africa Media Development Initiative.
43 Material for Africa also based upon the following sources: Calderisi, Robert, *The Trouble with Africa* (New York, NY: Palgrave MacMillan, 2006); Khanna, Parag, *The Second World* (NY: Random House, 2008); Maathai, Wangari, *The Challenge For Africa* (New York, NY: Pantheon Books, 2009); Mahajan, Vijay, *Africa Rising: How 900 Million African Consumers Offer More Than You Think* (Upper Saddle River, NJ: Wharton School Publishing, 2009); http://www.state.gov/r/pa/ei/bgn/5431.htm
44 Green, Matthew, "Downturn Hastens Nigeria's 'Brain Gain'," *Financial Times*, August 21, 2009, p. 4.
45 The World Bank Group, op. cit., p. 4.
46 The initial Yamoussoukro Treaty was signed in 1988 to liberalize aviation. However, it was not formally ratified until 1999.
47 International Air Transport Association, *Future Challenges in African Air Transport*, published by IATA, June 1985, p. iii.
48 Moores, Victoria, "Fighting Spirit," *Airline Business*, April 2009, p. 36.
49 Haruna, Waziri, "Is Nigeria Well on its Way to Being a Failed State?...No," *Focus on Africa*, July-September 2009, p. 33.
50 "Change for Good," *Airport World*, June-July 2009, Volume 14, Issue 3, pp. 19–20.
51 Green, Matthew, "The Making of a Megacity," *Financial Times*, 9 August, 2009, Arts and Leisure section, p. 1.
52 Ibid.
53 Material for Latin America based upon the following sources: Goodwin, Paul B., *Latin America* (New York, NY: McGraw-Hill, 2009) and Khanna, Parag, *The Second World* (NY: Random House, 2008).
54 According to Transparency International's 2005 "Corruption Perceptions Index."

55 van Agtmael, Antoine, *The Emerging Markets Century: How a New Breed of World-Class Companies Is Overtaking the World* (NY: Free Press, 2007), pp. 42–4.

56 "Latin American Aviation: Can Avianca Set the Synergy Flowing?'" *Centre for Asia Pacific Aviation*, 14 September 2009.

57 Smith, Geri, "Brazil's Coming Rebound," *BusinessWeek*, August 17, 2009, p. 42.

58 Ibid., pp. 42–5.

59 Müller, C., and Santana, E.S.M., "Analysis of Flight-operating Costs and Delays: The São Paulo Terminal Maneuvering Area," *Journal of Air Transport Management*, Volume 14, Issue 6, November 2008, pp. 293–6.

60 Smith, op. cit., p. 44.

61 Lima da Silva, Jefferson, "Azul: David Neeleman's New Blue Mission," *Airliners*, May/June 2009, p. 35.

62 Jana, Reena, "South America's New IT Hub: U.S. Tech Companies are Flocking to Chile, with its Savvy Engineers, Low Wages, and Government Grants," *BusinessWeek*, August 24 and 31, 2009, p. 73.

63 Associación Latinoamericana de Transporte Aéreo, Anuario 2008 Yearbook, p. 16.

Chapter 7

Flying with Tailwinds against Headwinds

The business model of the traditional legacy carriers (Aer Lingus, Japan Air Lines, US Airways, for example), developed mainly around "one-size-fits-all," is clearly ageing. Examples of the "one-size-fits-all" include hub-and-spoke systems, multi-class service, multiple types of airplanes, and later, membership in strategic alliances, as well as a new distribution channel, the airline website. In North America and Europe, almost no airline (with the possible exception of Lufthansa) managed to earn sufficient money over the last business cycle. Additionally, in many cases, consumer ratings are poor at best, and the employees do not appear to be enthusiastic about their deteriorating work rules, pension plans, and salaries. Similarly, the business model of the older-generation, low-cost airlines is also ageing, for both the well-established airlines, such as Southwest, and the high-profile airlines, such as Ryanair. In the case of the latter, for example, the business model continues to be based around the "rock bottom" cost structure. It has not been tweaked to generate any "wow" effect with either the carrier's traditional customers or to attract the cost conscious business travelers.

On the other hand, the business model of the newer-generation airlines, both the low-cost carriers (such as Air Arabia based in Sharjah, AirAsia based in Kuala Lumpur, and Spring Airlines based in Shanghai) or the new network airlines providing full service (such as Emirates, Etihad, and Qatar) is proving to be economically viable. Likewise, the revitalized business model of some older network carriers based in developing regions (such as Ethiopian, LAN, TACA, and Turkish) is enabling them to expand, and, for some, to even become global players (for example,

Ethiopian and LAN). Consequently, within the fundamentally and structurally changing environment (the headwinds), the "one-size-fits-all" business model cannot be expected to work for either all full-service, network airlines or for all low-cost, limited service airlines. Each airline within each sector must find its own viable business model. Furthermore, for those airlines who want to go beyond just searching for survival strategies, the key facilitators for re-vitalizing the business model are game-changing technology (relating to aircraft, IT, as well as business intelligence), and leadership to initiate innovation during times of change (tail winds). While game-changing technology can certainly facilitate the development of viable business models, it is game-changing leadership that can leverage technology to exploit the chaos, by either playing a different game, or by changing the rules of the existing game.

Game-Changing Technologies

Transforming Aircraft Technologies to Right-Size and Re-Shape Networks

As mentioned in Chapter 4, advanced airplane technology has always had a major impact on the global airline industry, as exemplified by the experience of the Airbus 330/Boeing 767 and the Bombardier CRJ/Embraer ERJ. This trend is expected to continue with new products not only from existing airplane manufacturers, but also from new manufacturers, such as the AVIC I Commercial Aircraft Company, Mitsubishi, and Sukhoi. Some of these new manufacturers have new technology airplanes on their drawing boards, while others are ready for flight testing. This section offers a scenario of the potential impact on airline operations of just two types of new airplanes, the long-range, wide-body Boeing 787 (and, of course its Airbus competitor A350XWB), and the standard body, Bombardier CSeries. Both airplanes are being designed to fly at the unit cost of larger airplanes in their respective categories. In the case of the Boeing 787, it is expected to match the seat-mile costs of the large airplanes such as the Boeing 777 and even the larger Airbus 380 (see Figure 7.1). Similarly, the

CSeries is being designed to match the economics of the current standard-body workhorses, such as the Airbus 320 and Boeing 737 (see Figure 7.2). (In both Figures, 7.1 and 7.2, the information shown is for illustrative purposes only.) Such economics, coupled with the range of these new products and potential liberalization policies of governments, have the potential to produce a dramatic impact on airlines' yields and networks worldwide.

Consider, first, the new long-range airplanes. Let us call the world's major, intercontinental, hub airports the "A" airports (measured in terms of not just the traffic volumes, but also the composition of the traffic). Hub airports in the "A" category would be Frankfurt, Heathrow, Hong Kong, Los Angeles, Narita, O'Hare, and Singapore. For the most part, these airports are not only connected through nonstop service, but they are also connected on a nonstop basis with more than 50 airports on other continents. Let us call the next tier down the "B" airports. This category would include such airports as Barcelona, Beijing, Boston, Johannesburg, Munich, and Seattle. Some of the airports in the "B" category are connected to "A" airports with nonstop

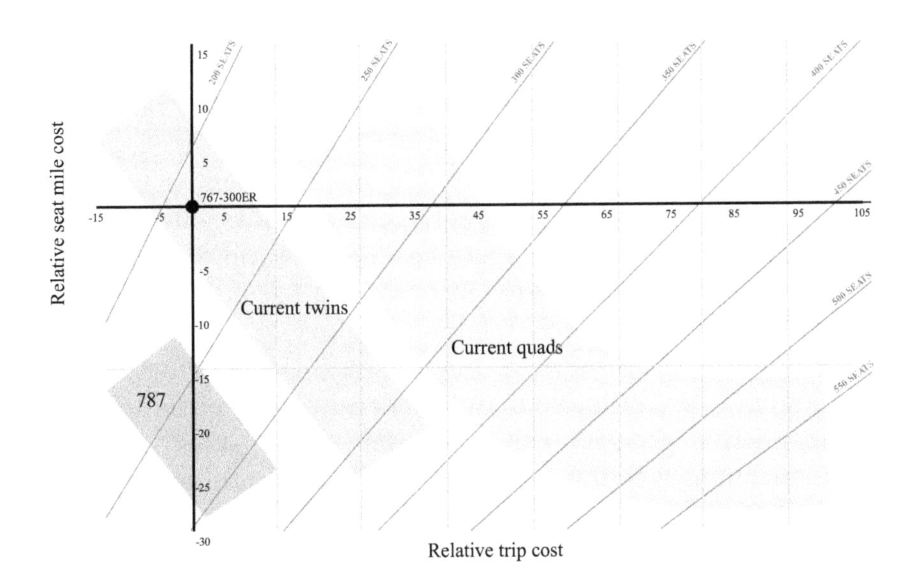

Figure 7.1 Relative Trip Cost vs. Relative Seat-Mile Cost
Source: Boeing Commercial Airplanes.

service (such as Seattle to Narita), and some are connected on a one-stop or connection basis, such as Boston to Singapore, and Manchester (UK) to Singapore. Let us call the next tier down the "C" airports. Examples would include Copenhagen, Las Vegas, and Manchester. Airports in this category would have very limited direct flights in intercontinental markets. For example, there is limited, nonstop service from the US to Manchester (UK), and from Las Vegas to London. How could the Boeing 787 change the network of some global airlines?

- More frequencies: Markets that are served only once a day with larger airplanes could increase frequency (for example, two flights a day), representing an enormous schedule improvement for business travelers and opening up more connectivity options at the hub airports.

- New routes out of existing hubs: "A" category hubs could easily be connected to many more "B" category hubs on a nonstop basis. For example, there could be a nonstop flight between Boston and Hong Kong. "A" hubs could also add nonstop connections to some of the current "C" category airports (making them "B" airports). This change would mean that many passengers could bypass "A" hubs. Just as a passenger in Seattle could fly, nonstop, to Singapore (without going through an "A" hub like Los Angeles), so could a passenger in Copenhagen, without going through Frankfurt. "C" airports, like Las Vegas, Lyon, Phoenix, Salt Lake City, St. Petersburg (Russia), and Stuttgart, could gain direct access between a larger number of intercontinental city pairs. Even more dramatic would be the emergence of many more destinations in India and China, provided their respective governments promote "open skies" policies. Such changes in the network could have a significant impact on the traffic feeding today's large hubs from "B" and "C" airports.

- Upgrade of "B" hubs to "A" level: As the ability to fly efficient, long-range routes will be increased with the new airplanes, "B" hubs can then be directly connected with numerous other "B" airports around the world. Boston, for example, could easily have nonstop service to Barcelona, Beijing, Johannesburg, and Munich. In some ways, "B" airports become "A" airports. In this scenario, the "B" airports will be the biggest winners resulting from the introduction of the Boeing 787. Losers will be the "A" airports, because of the vulnerability of their biggest feeder airports. The growth of the "B" airports will be further promoted by the CSeries planes, as explained below.

- Higher average yield on existing routes: Today's large, long-range airplanes usually carry many passengers with very low fares, because the seats are available and the marginal cost of an extra passenger is low. It is not unreasonable to assume that approximately 30 percent of today's long-range passengers pay fares that are not economically viable to fill a whole plane. The advanced-technology airplanes will enable airlines to "truncate the economy cabin" without sacrificing frequencies for the airlines' higher-yield passengers, resulting in a much higher average yield. Passengers traveling on the "truncated seats" may end up being transported on the higher capacity airplanes operated by the sixth-freedom carriers, further diluting their yields.

- Higher average yield on new routes: With new, nonstop routes, airlines do not have to share the revenue with airlines operating feeder flights, thus raising the average yield for the long-haul operator.

The combination of the introduction of smaller, advanced-technology airplanes and the proliferation of the "open skies" between major regions will provide an enormous opportunity for progressive airlines, who are willing to move fast, to optimize their multi-hub systems. Older-generation airlines with just one hub, no domestic traffic, and relying heavily on connecting traffic from "B" and "C" airports, could become losers. Moreover, airlines quickly adopting the capabilities of these advanced-technology airplanes will have other advantages. They will be able to provide service on new routes before the competition, and they will be able to reduce capacity quickly at appropriate times by downsizing from bigger airplanes, reducing frequencies, or by giving up "C" destinations.

Although on a different scale, the effects of the new CSeries airplane on the regional and intra-continental traffic will be similar. These airplanes will also enable airlines to increase frequencies, offer a greater number of nonstop services, and provide the opportunity to connect many tertiary airports. For example, passengers could be offered nonstop service between Boston and San Diego, between Manchester (UK) and Milan, and between Istanbul and Nairobi. As the new aircraft lower the profitability threshold for continental routes by about one-third, they support the upgrade of "B" airports to "A" level.

Next, small, advanced-technology airplanes (such as the CSeries) could have an enormous impact on the well-founded, regional airlines. Today, there are very successful regional (or supplemental) carriers, such as Skywest in the US and Flybe in Europe. They operate their fleets of 70- to 80-seat aircraft with "below scope clause" cost structures, either on behalf of major

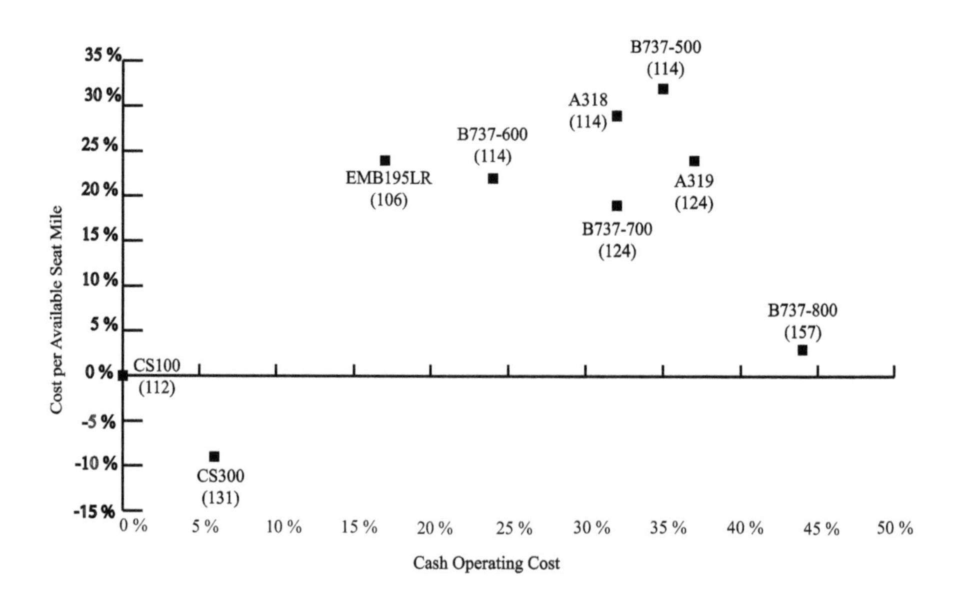

Figure 7.2 Cash Operating Costs vs. Cost per Available Seat Mile

Source: Bombardier Commercial Aircraft.

carriers, or with their own network. If these carriers are able to maintain their cost structure by flying the CSeries aircraft, they could become very aggressive intruders in the current low-cost arena, as they will be able to preempt some of the incumbent low-cost carriers from developing new routes. Such initiatives could dry up the growth potential of low-cost carriers in new markets, and re-direct them into competition with today's legacy carriers. One way to prevent the aforementioned scenario would be for the incumbent, low-cost carriers to use the platform of today's regional carriers to develop new routes with an "outsourced" CSeries operation (assuming they can negotiate their union contracts to allow this to happen), or the network carriers are able to attack the low-cost carriers with "on behalf" flights (again, under the provision of union consent to raise the scope clause threshold to more than 100 seats).

The availability of advanced-technology airplanes and advanced business analytics should help airlines make considerably better fleet decisions, based on criteria that go well beyond the standard net present value techniques. Here are just two examples. First, an airline could draw an insightful fan chart that shows the optimal point relating to an airplane capacity and its seat-abreast configuration. With the older generation standard-body aircraft, four-abreast seats might be more economical for aircraft capacity under 70 seats. Five-abreast might be more economical for about 100 seats and six-abreast for standard-body aircraft with 140 or more seats. However, new technology five-abreast standard-body aircraft have the potential to be economically viable for seats that approach 140. Second, advanced business analytics should be used to evaluate the performance and incremental capacity of new airplanes with respect to their cost and value. The standard fan diagrams can include not just the cost per trip vs. cost per seat mile, but also the revenue per trip vs. the revenue per seat mile. Such a diagram would then portray airplanes based on profit per trip vs. profit per seat mile. Moreover, the analysis can be performed at an individual route level as opposed to a given set of markets. Even for individual markets, the analysis can be performed on a point-to-point nonstop basis vs. a mixture of point-to-point and connecting traffic, not to mention the type of connecting traffic. The key points, however, are that (a) if the business analytics are developed at the profit per trip vs. profit for seat mile, then it becomes necessary to include competitor strategy and reaction, and (b) the fleet analysis must be performed in an integrated manner, not a sequential manner.

Finally, another effect of both airplanes entering the market could be a dramatic drop in the residual values of today's larger airplanes. In order to compete with the low operating cost of the new airplanes, the capital cost part of the incumbent planes has to shrink. If Bombardier were to succeed in selling thousands of their CSeries airplanes, the values of half time Airbus 320s and Boeing 737s will have to be reassessed. This result will undoubtedly cause serious concerns for the large leasing companies who currently are the main supply sources of these airplanes.

Transforming Information and Communication Technologies to Improve Customer Experience and Operational Efficiency

Just as emerging aircraft technologies can enable airlines to re-size and re-shape their networks, there is a set of emerging Information and Communication Technologies that can enable airlines to improve their customer service and their operational efficiency. SITA lists the following technologies in its New Frontiers Paper.[1]

- Mobile Devices;
- Web 2.0;
- Near Field Communication;
- Radio Frequency Identification;
- Biometrics;
- Service-Oriented Architecture;
- Collaborative Decision Making;
- Cloud Computing;

- Visualization.

In this paper, SITA discusses the above nine technologies in two parts. Part I encompasses technologies that will benefit passengers' journeys, improving their travel experience. Part II discusses technologies that will improve the operations of the air transport industry, leading to an improvement in profitability. The role of Radio Frequency Identification (RFID) is discussed in both parts.

Within the first category, mobile devices are not only blurring the distinction between phones and computers, but they will also enable much greater access to air travel services while passengers are on the move. Coupled with the emerging Web 2.0 capability, mobile devices can provide travelers with relevant, personalized, and timely information. They can also replace the boarding passes currently being printed by check-in kiosks or home computers. This capability is already being deployed by airlines such as Lufthansa. The near field communication (short-range, high-frequency wireless communication) can enable passengers to use their mobile, hand-held devices to check-in themselves and make payments. The Radio Frequency Identification, enabling the storage of a large amount of personal data on chips, can bring about vast improvements in the processing of passengers at airports and the tracking of luggage. Similarly, advances in biometrics technology can lead to substantial improvements in processing passengers while achieving higher levels of security.

Technology advances within the second category relate to open standards that work "behind the scenes" to produce benefits for the providers of air transport services and, in turn, for the users of air transport services. The Service-Oriented Architecture (SOA) enables the IT department of an airline to develop applications more rapidly, enabling the airline to have greater business agility. Technology advances in collaborative decision-making relate to the ability to share information at an exponential rate, leading to an improvement in the decision-making process. Cloud computing refers to the ability of an airline to meet its computer needs (hardware, software, and data storage) via the Internet. This technological advance can enable an airline's IT department to access applications and services, reduce the IT department's capital costs, and enable it to pay on a per-usage basis. Virtualization can enable an airline to share its server capacity within the entire organization, including facilities that are established at different locations. This technological advance enables large airlines to optimize their IT assets and provide greater protection against failures or overloads at one location. At the same time, it reduces the barriers of entry for new airlines, as building up an IT infrastructure around new technologies may be cheaper and faster than adapting legacy applications to the new standards on a piecemeal basis.

Game-Changing Leadership

The current chaos is clearly distinguishing two groups of airline leadership. One group is looking for strategies to survive by reducing their areas of vulnerability. That is what this group will achieve, survival and only survival, possibly with the help of their governments. This group does not include just airlines from developing economies. Consider, Japan Airlines, the largest carrier based in the Asia-Pacific region, and in the world's second largest economy. Even though it was privatized in 1987, it has been receiving government support since 2001, with over a billion US dollar loan earlier in 2009 that was guaranteed by the government.[2]

The other group is looking to exploit chaos through game-changing strategies. Consider a couple of examples. First, within different sectors of the airline industry, while the legacy airlines have been shrinking, the low-cost airlines have been expanding, although maybe at a lower rate than before. This phenomenon is visible in all regions of the world, as exemplified by the operations of WestJet in Canada, Azul in Brazil, Air Arabia in the Middle East, AirAsia in Malaysia, Spring Airlines in China, and V Australia. Some of their growth is coming from the stimulation of traffic from the introduction of low fares. Some is coming from the trading down phenomenon, resulting from the economic downturn. Another part is coming from the development of hybrid airlines that offer leisure travelers low fares and business travelers frequent schedules and connectivity at lower fares. Examples of these hybrid airlines are AirAsia X, Tiger Airways, and Virgin Blue. AirAsia X has announced its decision to introduce Business Class seats on its A340s, estimating that they could capture 15 percent of the corporate business.[3] They are likely to succeed where others, such as Oasis Hong Kong, failed because of its lack of higher levels of frequency, feeder traffic, brand extension, and operating scale.

Next, consider airlines that operate within the same sector. The economic downturn that began at the end of 2007 (coupled with the high price of fuel, the expansion of low-cost carriers, the strength of the Japanese Yen, and the impact of the H1N1 flu virus) had a negative impact on the financial performance of both Japan Airlines and All Nippon Airways. Both needed to raise funds. As reported in the media, Japan Airlines raised funds from private sources (with most of the funds provided and guaranteed by the government) to cover its hemorrhaging losses. All Nippon, on the other hand, raised capital from private sources on its own, and is reported to be using the funds raised to finance its growth plans, enabled by the enormous increase in airport capacity (both at Narita and Haneda).[4]

Consider two other cases, South African Airways and Ethiopian Airlines. South African Airways, a dominant carrier in Africa, was struggling even before the economic downturn that began in late 2007. Its poor financial performance led the airline to ask its government for financial support to facilitate its restructuring initiatives (just to survive). Despite the government support, the airline has not been able to develop a viable business model, either to defend its share of the domestic markets or its presence in international markets. Seeing this weakness and the strength of its geographic location, Ethiopian Airlines is expanding its services through its Addis Ababa hub. The carrier currently serves 33 points in Africa and 14 in the Gulf, Middle East and Asia.[5] The long-haul, wide-body aircraft on order (such as the Boeing 787 discussed previously) will enable Ethiopian to expand its services to not only Asia (to capitalize on the growing traffic between Africa and China), but also to North America (where the airline currently has one-stop service to Washington).

As mentioned throughout the book, airlines will be facing a new world. They can either just respond to it through defensive strategies, or they can embrace it and go for offensive strategies. Imagine the following scenario. Figure 7.3 shows the market capitalization of a couple of dozen airlines based in different parts of the world. Consider the level to which Ryanair has already risen. Within the group of airlines selected, as of June 2009, Ryanair had a higher market value than Delta in the US, Lufthansa in Europe, and Cathay in Asia. The media has reported a number of times the possibility of Ryanair serving intercontinental markets, such as between the US and Europe. While numerous other upstarts did not succeed, Ryainair has the resources to succeed. Leaving aside its lower operating costs and its unconventional strategies, just consider the enormity of its feeder traffic. If Ryanair were to start service from, say, Stansted airport in the greater London area, it could, potentially, have feeder traffic from 13 cities in Germany, 20 cities in Italy, 14 cities in France, and 17 cities in Spain. This feeder traffic could increase exponentially if Ryanair were to start service to just four cities across the Atlantic, such as one near Boston, one near New York, one near Washington, and one near Toronto (see Figure 7.4). Very quickly, Ryanair could expand its service to more than a dozen, metropolitan areas of the US and Canada. Are the major legacy carriers on both sides of the Atlantic prepared for such a scenario? Should they develop defensive or offensive strategies?

So, how does an airline embrace chaos and exploit it? The key is to face the new reality and re-establish priorities in line with the new reality. Let us take the point about facing reality. First, there are now viable new competitors. British Airways is now competing not just against Singapore, but also Emirates and AirAsia X. In the case of Lufthansa, the fierce competitors in their future may not be just British Airways-Iberia and Air France-KLM, but also Ryanair, easyJet and Air Berlin. Those are just the potential airline competitors. What about potential format invaders from outside the airline industry that could change the role of an airline to be simply a provider of air transportation service between airports?

Second, the government intervention is not going away to produce a level playing field. Governments will still control, for example, airlines' access to global capital, their ability to consolidate across borders, and the facilitation processes at

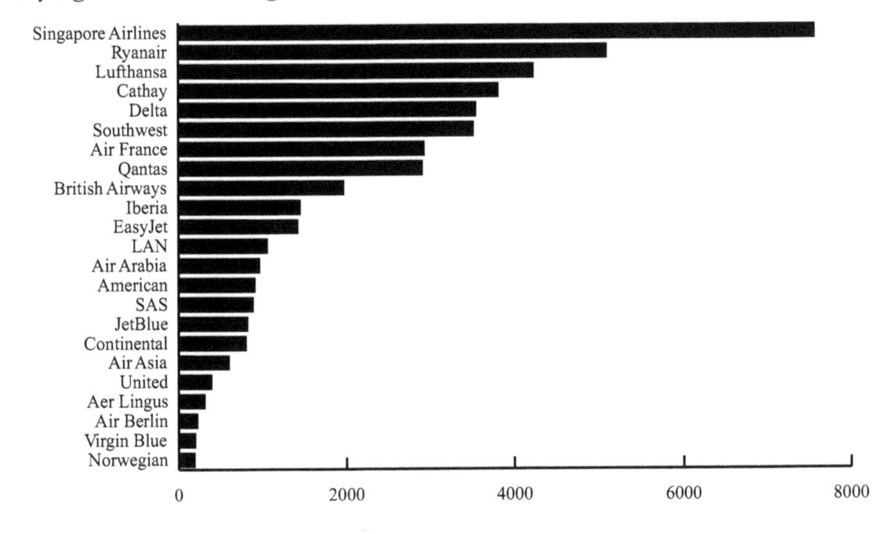

Figure 7.3 Airline Market Capitalization (€M), June 2009
Source: Aviation Strategy.

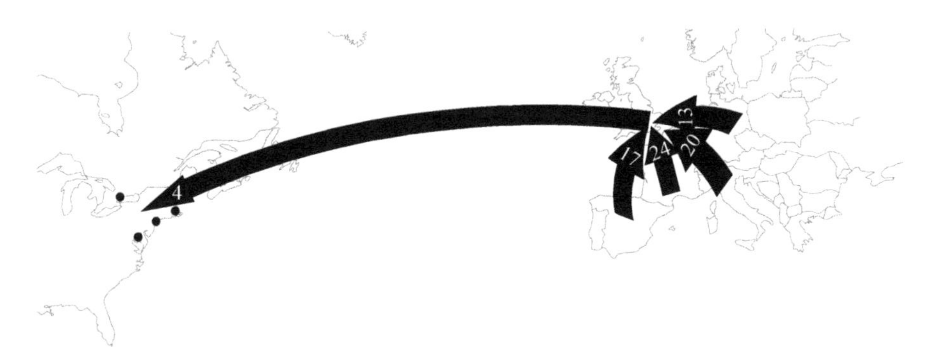

Figure 7.4 Potential Opportunities for Transatlantic Service by Ryanair
Source: Data derived from the route map shown on the website.

airports. Third, the center of gravity with respect to the aviation industry is moving. As shown in Figure 7.5, market share of airlines based in different regions of the world has been changing for some time, but the rate of change has increased. The global market share held by US airlines is forecast to further decrease substantially, while the market share of Asian airlines is expected to increase substantially. Similarly, European airlines are expected to lose market share while airlines in Africa, Latin America, and the Middle East increase their market shares. The forecasts extending to the year 2030 are based on the capacity ordered by various airlines based in each region.[6] Finally, consumer behavior is changing rapidly with respect to consumers' buying patterns (it is not just trading down that may be a short-term trend, but consumers buying smartly that will be a long-term trend) and their expectations (better customer experience, and more control, as discussed in Chapter 3). Accordingly, consider the possibility of one of the IT giants using its Web 2.0 databases and communities to become an "online travel forwarder" to meet the needs of the changing demand of a new generation of travelers.

Many airlines are still in denial about the severity and the agility needed to deal with this chaos, and they continue to point to the inherent constraints within the airline industry, limiting their ability to adapt to the marketplace on a timely basis. As Chapter 2 showed, such a denial about facing the realities of the marketplace can lead to the downfall of a once mighty company, such as General Motors. Consequently, in short, competition will come from dynamic new business model carriers within the region, old business model carriers from more dynamic regions, and from format invaders from more dynamic industries, such as IT, communications, and consumer lifestyle brands.

Once managements begin to face realities, the next step is to re-establish priorities. If one accepts that the playing field is not going to become level in the foreseeable future and that constraints inherent within the airline industry (relating to infrastructure, labor, the nature of operations, for example) are not going away, then airlines must find ways to innovate around these realities. Similarly, priorities must change to develop products that meet the needs of contemporary customers. As illustrated in Chapter 5, the auto manufacturer Hyundai introduced a new value proposition to cater to changing buyer behavior. While most automakers were focusing on reducing prices dramatically, Hyundai announced an alternate value proposition, one that had nothing to do with price. Consumers responded favorably to the new value proposition.

A critical success factor relating to the re-establishment of priorities relates to the execution of new business models

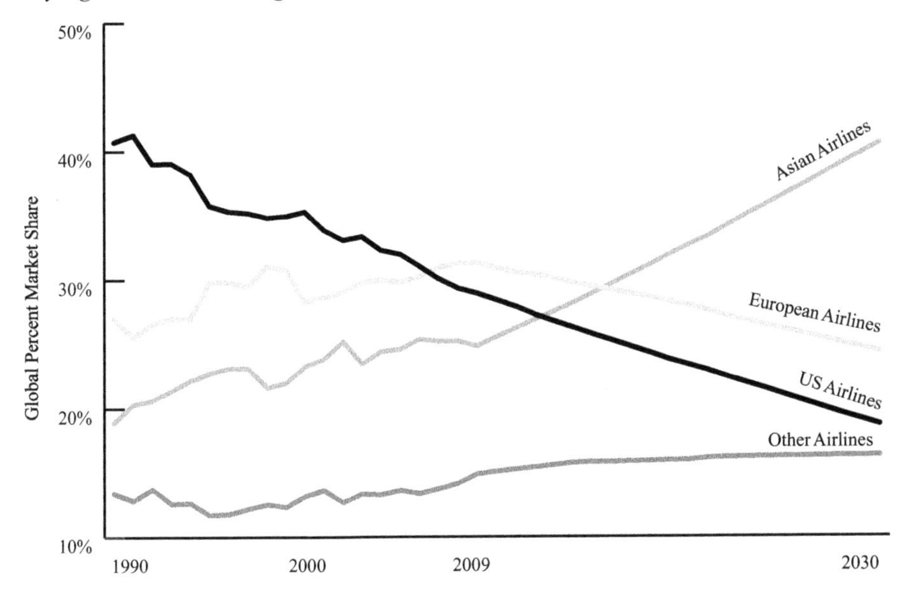

Figure 7.5 Percentage Share of (Global) Traffic (RPMs)
Source: The Airline Monitor, July 2009, p. 40.

developed to adapt to the marketplace. In the past, even airlines that did try to change their business models, did not implement the changes well. Recall the number of legacy carriers that set up their low-cost carriers and did not succeed, as exemplified by the experience of British Airways, Continental, Delta, KLM, SAS, United, and US Airways. Recall, also, the experience of a successful low-cost carrier buying a legacy carrier to operate a two-brand airline (GOL acquiring Varig). So far, only a limited number of airlines have been successful with the multi-brand strategy. Examples include Air France-KLM, LAN, Lufthansa-Swiss, Qantas, and TACA. In these cases, innovation was achieved more from the execution of the concept than the concept itself. The key component of innovation relating to execution has been management's ability to deploy the new business model alongside the existing model. As for working side-by-side, Jetstar started to serve Narita, replacing Qantas. The successful execution of a multi-brand model can pay off in good times and in bad times, as exemplified by the experience of Qantas. It is Jetstar that could be the operator of most of the Boeing 787s ordered by Qantas.[7]

Closing Thoughts

Because the airline industry faces regulatory, political, economic, and operating constraints, past and current managements have necessarily been more concerned with coping and compromising than they could have been with achieving any kind of significant breakthroughs and progress. It is not the case that they have not had the desire and interest to do so. If airline managements were able to operate and compete in a freer and more open marketplace, of course, how airlines are managed would forever be changed. Weak carriers would have disappeared faster, and strong carriers would have been rewarded faster for their innovations, thus giving the traveling public much better service. However, most governments around the world are not willing to let their weak airlines go out of business.

The realities of the marketplace, however, are not going to change in the foreseeable future. Therefore, airline managements must innovate around the realities in the global marketplace, and incorporate best global business practices to the extent they can. As hypothesized in the first chapter, airlines can expect to operate within a new normal. Moreover, as indicated in the previous chapter, airlines can expect "pockets of growth" around the world, centered in metropolitan areas. Consumers in all these metropolitan areas will expect to be connected, more or less directly. Consider how TAP Portugal has been able to exploit its geographic location and a different mindset to establish an "outside hub" to serve the metropolitan areas of Brazil (see Figure 4.2), how Emirates has been able to bundle traffic from India and Australia to the UK and Europe via Dubai (see Figure 1.4), and how LAN has been able to get around the standard ownership and control regulations to develop one brand and four production units to increase aircraft and crew productivity (see Figure 4.3). IT technology (of the type discussed in Chapter 4) and new airplanes on the horizon (discussed previously) can help in the development of a viable network as well as product and services to meet the needs of emerging segments of consumers. It is now up to the leadership as to whether they want to develop strategies simply to survive the chaos, or to exploit it. Since this book points to insights from other businesses (relating to both failures and successes), let us end with two thoughts. First, over the years, more than a dozen prominent businesses were started during recessions. Here is a list of just six of the many: CNN, Federal Express, General Electric, Hewlett Packard, Hyatt, and Microsoft.[8] Second, chaos presents opportunities for entrepreneurs and format invaders. If Apple changed the way how people hear, buy, and make music,[9] is there a format invader waiting to transform the way people plan, buy, and experience air travel?

Notes

1 Peters, Jim, "Ten Technology Advances that Will Change Air Travel," *A New Frontiers Paper*, SITA, 2009.
2 Flottau, Jens, and Bradley Perrett, "Calling Tokyo: Japan Airlines Seeks a Foreign Strategic Investor—and not Necessarily from Oneworld," *Aviation Week & Space Technology*, September 21, 2009, pp. 34–5.
3 Ballantyne, Tom, "Boom Time for Budget Carriers," *Orient Aviation*, September 2009, p. 16.
4 Tudor, Geoffrey, "ANA Posing Threat to JAL," *Orient Aviation*, September 2009, p. 36.
5 "South African Airways and Africa's Airlines: Outlook Bleak for SAA, " *Centre for Asia Pacific Aviation*, 27 August 2009.
6 *The Airline Monitor*, July 2009, pp. 32–40.
7 "Low Cost Airlines: The World has Changed," *Centre for Asia Pacific Aviation*, September 2009.
8 Caron, Sarah, "14 Big Businesses That Started on a Recession," *Inside CRM*, November 11, 2008. http:/www.insidecrm.com/features/businesses-started-slump-111108/
9 "Music to Their Ears," *Fortune*, November 23, 2009, p. 106.

Index

About the Author

Nawal Taneja's career in the global airline industry spans 40 years. As a practitioner, he has worked for and advised major airlines and airline-related businesses worldwide, facilitating their strategies. His experience also includes the presidency of a small airline that provided schedule and charter services with jet aircraft, and the presidency of a research organization that provided advisory services to the air transportation community worldwide. In academia, he has served as Professor and Chairman of Aerospace Engineering and Aviation Department at the Ohio State University, and an Associate Professor in the Department of Aeronautics and Astronautics of the Massachusetts Institute of Technology. On the government side, he has advised civil aviation authorities in public policy areas such as airline liberalization, air transportation bilateral and multilateral agreements, and the financing, management, and operations of government-supported airlines. He has also served on the board of both public and private organizations.

He holds a Bachelor's degree in Aeronautical Engineering (First Class Honors) from the University of London, a Master's degree in Flight Transportation from MIT, a Master's degree in Business Administration from MIT's Sloan School of Management, and a Doctorate in Air Transportation from the University of London. He has authored five other books for practitioners in the airline industry: (1) *Driving Airline Business Strategies through Emerging Technology* (2002); (2) *AIRLINE SURVIVAL KIT: Breaking Out of the Zero Profit Game* (2003); (3) *Simpli-Flying: Optimizing the Airline Business Model* (2004); (4) *FASTEN YOUR SEATBELT: The Passenger is Flying the Plane* (2005); and (5) *Flying Ahead of the Airplane* (2008). All five books were published by the Ashgate Publishing Company in the UK.

For Product Safety Concerns and Information please contact our EU
representative GPSR@taylorandfrancis.com
Taylor & Francis Verlag GmbH, Kaufingerstraße 24, 80331 München, Germany

www.ingramcontent.com/pod-product-compliance
Ingram Content Group UK Ltd.
Pitfield, Milton Keynes, MK11 3LW, UK
UKHW021834240425
457818UK00006B/194

* 9 7 8 1 4 0 9 4 0 0 9 9 8 *